Principles of engineering organization

ENGINEERING MANAGEMENT

Series Editor
S. H. Wearne, PhD, BSc(Eng), DIC, CEng, FAPM

Editorial Panel
D. Neale, CEng, FICE, MIHT, MBIM
D. P. Maguire, BSc, CEng, FICE
J. V. Tagg, CEng, FICE, FIStructE, FRICS, FCIArb
J. Bircumshaw, CEng, MICE
M. H. Denley, BSc, CEng, MICE
D. G. Cottam, BSc, CEng, FICE

Other titles in the series
Civil engineering insurance and bonding, P. Madge
Marketing of engineering services, J. B. H. Scanlon
Managing people, A. S. Martin and F. Grover (Eds)
Civil engineering contracts, S. H. Wearne
Control of engineering projects, S. H. Wearne (Ed)
Construction planning, R. H. Neale and D. E. Neale
Management of design offices, P. A. Rutter and A. S. Martin (Eds)
Financial control, N. M. L. Barnes (Ed)
Project evaluation, R. K. Corrie
Construction management in developing countries, R. K. Loraine

ENGINEERING MANAGEMENT

Principles of engineering organization

Stephen Wearne, BSc(Eng), DIC, PhD, FAPM

Thomas Telford, London

Published by Thomas Telford Services Ltd, Thomas Telford House, 1 Heron Quay, London E14 4JD

Distributors for Thomas Telford books are:
USA: American Society of Civil Engineers, Publications Sales Department, 345 East 47th Street, New York, NY 10017-2398
Japan: Maruzen Co Ltd, Book Department, 3-10 Nihonbashi 2-chome, Chuo-ku, Tokyo 103
Australia: DA Books and Journals, 11 Station Street, Mitcham 2131, Victoria

First edition published 1973 by Edward Arnold (Publishers) Ltd
Second edition published 1993

A catalogue record for this book is available from the British Library

Classification
Availability: Unrestricted
Content: Guidance based on best current practice and research
Status: Refereed
User: Engineers and engineering managers

ISBN: 0 7277 1656 5

© Stephen Wearne, 1973, 1993

All rights, including translation reserved. Except for fair copying, no part of this publication may be reproduced, stored in a retrieval system or transmitted in any form or by any means, electronic, mechanical, photocopying or otherwise, without the prior written permission of the Publications Manager, Publications Division, Thomas Telford Services Ltd, Thomas Telford House, 1 Heron Quay, London E14 4JD.

The guide is published on the understanding that the authors are solely responsible for the statements made and opinions expressed in it and that its publication does not necessarily imply that such statements and or opinions are or reflect the views or opinions of the publishers. Every effort has been made to ensure that the statements made and the opinions expressed in this publication provide a safe and accurate guide; however, no liability or responsibility of any kind can be accepted in this respect by the publishers or the authors.

Typeset in Great Britain by MHL Typesetting Ltd, Coventry
Printed and bound in Great Britain by Halstan & Co. Ltd, Amersham, Bucks.

Dedication

This book is dedicated to the memory of
G. R. (Ron) Thomas, CEng, MIEE, FBIM, FIIM

Preface

Organization is a means of enabling people to achieve more together than they could achieve alone; or it should be.

This book has been written for engineers and managers, to assist them in understanding and improving systems of organization. The book provides a review of principles and some analysis of examples drawn from a range of engineering activities.

A standard system of organization that suits all demands has not yet been evolved. Probably it never will be, because firms vary in their objectives, culture, work and circumstances. Principles to apply in analysing or in designing a system are considered in this book. Alternatives are reviewed, and their potential advantages and disadvantages are compared.

The examples indicate the variety of systems to be found in practice. Common to all of them is the need to think about how and the extent to which people's knowledge, interests and skills can be best organized so as to achieve more than they could achieve randomly or out of habit, trial and error.

Part I of this book consists of chapters on the needs of engineering projects and systems of organization. The first chapter introduces the principle of designing an organization to suit the work that is its primary task. The work typical of any engineering project is considered in chapter 2. A cycle of activities common to new products, structures and other projects large and small is depicted. Chapter 3 goes on to consider the pattern of decisions required for a project.

Part II consists of chapters on the relationships between engineering firms and their customers in various industries. Chapter 4 is the first of these. In it, the pattern of specialization by firms in manufacturing is considered. Chapter 5 describes the contrasting pattern of specialization in construction. Readers unfamiliar with the construction and manufacturing industries should read chapters 4 and 5 in order to appreciate the differences between these industries and to be given the basis for considering the ways in which systems of organization have evolved within firms under these different conditions. Chapter 6 adds some observations on joint working by firms.

Part III consists of chapters on the choices in organizing engineering work among the people in a firm. The principles seem elementary when considered one at a time. Their combination to meet varying demands and projects differing in size, programme, location, etc., can lead to quite complex systems. Examples from companies and public authorities are described in chapter 7. The choices of system are reviewed in chapter 8, and the coordination and control of projects are reviewed in chapter 9. Chapter 10 gives some final comments.

Appendix A explains words and types of diagram used in describing systems of organization in firms. Appendix B summarizes choices in employing one or more specialist firms to supply part or all of a project. Appendix C sets out principles and procedures for defining the lessons of engineering projects in order to apply them to future ones. Appendix D lists the publications of authors quoted in the book. Suggestions for further reading are given at the end of the chapters.

This book is one result of continuing studies of theory and practice in engineering management begun in industry and developed at Manchester and Bradford.

These studies have depended on the time and interest of many firms and of the individuals in them who have helped us try to draw together ideas and the lessons of experience. Thanks are due to all of them, and equally to past and present teachers, employers, colleagues and other helpers at home and abroad. I am particularly grateful for comments and material from Mr R. Levy, Mr A. S. Martin and Mr R. D. Thomas.

The results as they appear in this book are, of course, the author's sole responsibility.

S. H. Wearne
Applegates
November 1992

Definitions used in this book

The word *organization* is used here to mean the condition of being formed 'into a whole with interdependent parts, to give a definite and orderly structure ...' (O.E.D.). It can apply to firms or to departments or groups within a firm. *System* is used to mean the way in which the work of groups of people in an organization fits together or is intended to fit together.

Firm is used to mean a company, a municipal or other public corporation, a privately-owned partnership or the equivalent of any of these such as a subsidiary of a firm or a jointly owned subsidiary. *Industry* is used to mean all who produce goods or services, privately or publicly owned. *Sector* of industry means all the firms who supply a type of product or service.

The word *project* is used to mean any new, replacement or improvement of a unit of industrial production or construction large or small, and to cover its creation from the first study of ideas to its completion to produce useful goods or services.

Engineering is the process of making decisions and of utilizing resources for a project.

The term *staff* is used to mean service roles in a 'line-and-staff' system of management, as discussed in Chapter 9. *Staff* is therefore *not* used here as a collective noun for the people who work in offices.

Contents

Part I. Engineering organizations and projects

1 The primary task   3
Specialization, size and communications; survival and success; organization practice and theory; systems and flexibility

2 Engineering projects   6
Projects; tempo; resources; activities for a project; stages of work; initial stages; selection; design, development and research; manufacture and construction; cancellation; information flow; feedback of information; interdependence of projects; feedforward to further projects

3 Engineering decisions   18
Decisions in design; decision sequences; transformation of information; critical decisions; innovation; development; classes of problems; decision processes; organizational consequences

Part II. Relationships between organizations

4 Manufacturing   29
Change and complexity; specialization of firms; the evolution of engineering; interdependence of firms; projects; project relationships; contracts; innovation; relationships between firms

5 Construction — 39
Construction conditions; specialization of firms; traditional contract relationships in Britain; promoter, engineer and contractor; specialization and continuity; authority–contractor relationships; the engineer in an authority; innovation; industrial projects; direct labour; comprehensive contracts; consultants; evolution of systems

6 Joint ventures and consortia — 49
Joint projects; definitions; horizontal collaboration; vertical collaboration; complex collaboration; complexity and control; integration of firms

Part III. Relationships within organizations

7 Case studies — 59
The cases; manufacturer **M**; manufacturer **N**; licensing; promoter **P**; consulting engineering firm **Q**; contractor **R**; comment; continued evolution

8 Specialization within firms — 97
Departments; grouping by project, subject, project stages, levels; functional grouping; drawing offices; separation for cost control

9 Coordination and control — 107
Managerial hierarchies; delegation of authority; external commitments; complexity and risks; deputies; functional supervision; separate coordinators of projects; line-and-staff roles; evolution of a system; matrix systems; relative responsibility; matrix management or internal contracts?; three dimensional matrix; numbers

10 Final comments — 127
Principles, practice and people; projects and organizational change; larger and more organic systems; data systems; uncertainty and safety; coordination; project teams; case studies; flow diagrams; variety and further studies

Appendices

A	Charts and diagrammatic conventions	135
B	Number and scope of contracts	142
C	Conduct of project completion reviews	146
D	Bibliography — Sources	150

PART I
Engineering organizations and projects

1 The primary task

Specialization, size and communications

As stated at the beginning, a system of organization should be a means of enabling people to achieve more together than they would randomly. Specialization by firms and by groups of people in them to develop expertise is valuable for achieving this objective, so is growth in the size of firms and their projects to gain the potential advantages of economy of scale.

Greater specialization and large organizations are characteristics of the evolution of civilization, not just in industry. Engineering is one example. Not all projects and organizations are large and complex. They vary greatly, but in most firms the effectiveness of design, development and other engineering activities has become dependent upon systematic relationships to link together work by two or more firms and specialist groups of people in them. Engineering managers should therefore be concerned not only with *what* are the answers to project problems and technical questions; they should also analyse *how* the problems arise and *how* the decisions on them are made.

Survival and success

An engineering organization is expected to give value in return for the resources it uses, whether it is part of a company established to trade profitably or whether it is in a publicly-financed authority established to provide a service. In other words, it survives and succeeds by transactions with its surrounding systems.

To design or analyse the system of organization in a firm one should therefore start by defining the external relationships needed to achieve what A. K. Rice and other observers of industry call its *primary task* — i.e. the task or set of tasks it must carry out to survive and succeed. To do so, the firm's dependence on the current and future demand for its products and services, and also on suppliers must be analysed, as in the chapters which form Part II of this book. Then the sequence of internal activities to carry out the primary task can be analysed and the relationships needed between the people employed on these activities defined, as in the chapters in Part III.

Organization practice and theory

Theories on how to organize people at work have long been an interest of managers, engineers and observers of industry. A pioneer in this was the Mancunian J. Slater Lewis,* whose book published in 1896 included the first known examples of a communications diagram and organization chart. Many observations and views have appeared since, particularly from the USA.

The result is a large quantity of 'organization theory'. More appears daily. Not all of this literature distinguishes between tested theory, logical ideas and personal impressions. All these can be valuable, but not equally so. Statements on an organization from any of its members are, for instance, likely to be influenced by personal interests. As such, their views cannot be assumed to be reliable. An independent observer should be more reliable, but the observations may not be valid. The ideas and examples in this book should therefore be assessed according to their source.

Much organizational practice has been based on the problems of large factories and bureaucracies. In these, efficiency has been pursued by dividing work into small repetitive tasks. This view of how to organize work has been very influential. Its impersonal practice of defining who is expected to do what can help to achieve the primary task, but only so far as the relationships needed can be predicted. The overcoming of unpredictable problems demands 'organic' relationships, as observed by T. Burns and G. M. Stalker. By organic is meant flexibility, informality and uncertainty in relationships. As observed in a recent review by D. Dill and A. W. Pearson, the least division of work is appropriate for engineering and other development organizations, in order to innovate, motivate people, be flexible in utilizing individuals, and avoid divisions of interests and objectives.

---

*See publication listed in Appendix D.

Systems and flexibility
The bureaucratic and manufacturing origins of much experience and theory can induce a tendency to try to establish a steady state of relationships in an organization, and to optimize costs, value and safety in using resources. This may be successful in a firm for a limited time. A steady state enables people to develop skills and routines, but the external demands are not steady and not all changes are predictable. Technology, social values and trading conditions in the world are all beyond the control of any one firm or industry. There is, for instance, a continuing pressure for better products and services, though it is not always agreed what is 'better', and projects vary in their timing, risks, size and importance. The result is that the demands on sectors of industry and individual organizations change, often requiring only adaptation but sometimes causing emergencies and discontinuities.

Bureaucratic habits and traditions have become strongly established in engineering over the past hundred years. They give a sense of internal order, but with the risk that the rules and routines displace attention to objectives. An organization needs an open system and the simplest appropriate for its primary task.

Further reading
Handy C. B., *Understanding organizations*, London, 1985, 3rd edn.
Kakabadse A., *Working in organizations*, Penguin, London, 1988.

2 Engineering projects

Projects
The primary task of engineers in industry varies in scale and complexity from small improvements to products to large capital investments, but the term *project* is becoming widely understood to mean any such investment of resources to achieve a distinct objective.

The distinction of a project compared with previous ones may be only in detail to suit a change in market or a new site. The distinction may also include some novelty in the product, in the system of production, or in the equipment and structures forming a system. Every new model of car, aircraft, ship, refrigerator, computer, crane, steel mill, refinery, production line, sewer, road, bridge, dock, dam, power station, control system, building or software package can be called a project. This is also the case with many smaller examples, and the characteristics of a project are also found in the work of replacing or altering a product or a service. The common use of the classification of project for any of the above is logical because all examples share these characteristics.

- A project depends on an *investment* of resources for an *objective*.
- A project causes irreversible *change*.

Tempo
To get value from the investment, the time allowed for completion of a project is usually limited. As a result, a project is a period of intense engineering and other activities, but is short in its duration relative to the subsequent working life of the investment. This is illustrated in Fig. 2A.

After the initial study of ideas for a project, the decision to

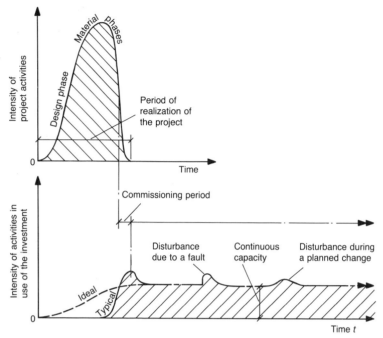

Fig. 2A. Relative tempos of a project and of its product in use

proceed leads to an increase in resources employed but this is followed by dispersal of these as the work is completed. The rate of concentrating resources on a project to hasten its completion may be limited by technical reasons, such as space, or whether or not there is a cash income to be obtained from the use of the project when completed.

As a project approaches completion, the importance of economy in the use of resources tends to become secondary to the expedients adopted to save or to recover time. The intensity of activities therefore increases. By then the people on the project may be learning how to work together effectively, but the value of this is limited because many activities are relatively brief. A rapid build-up is followed by a usually more rapid fall, as sketched in the upper part of Fig. 2A.

By contrast, the use of the completed project is a continuing activity, as sketched in the lower part of Fig. 2A. It differs from the project stages in nature and in tempo.

Extensions, improvements and planned maintenance to the completed facility during its use are, in effect, additional projects, thereby disturbing the continuity, as also indicated in Fig. 2A.

Resources

The demands for the engineering capacity and other resources required to create a project (or to alter one) are therefore transitory. To be economic, these resources must be used as continuously as possible on a series of projects. There is thus a potential conflict between two objectives

- achieving the sequence of activities essential to each project
- sharing resources among a variety of projects.

Continuity in the use of resources and the development of skills may not suit the order of their use that is ideal for a project. Analysis of the pattern of activities essential for projects should provide a basis for planning how to try to meet both these objectives.

Activities for a project

The projects in hand at any time in industry obviously differ greatly in their size, technical content, novelty, urgency and duration, but it is common for the work for every project to consist of a branching network of activities as depicted in Fig. 2B. The

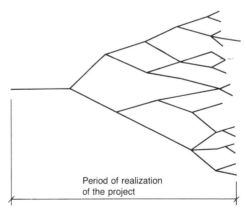

Fig. 2B. Typical pattern of activities through a project (the convention that time elapses from left to right in such diagrams is discussed in Appendix A)

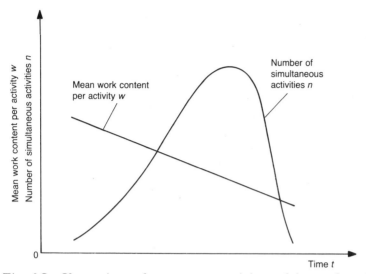

Fig. 2C. Change in work content per activity and in number of simultaneous activities during a project

sequence shown in this proceeds from a few general and abstract activities to many smaller and increasingly simultaneous ones. Every activity may be dependent on the completion of the prior ones in the sequence, so that the freedom of choice decreases as a project proceeds to the large scale of interdependent detail.

Techniques based on the network of activities have been developed to plan the design, material ordering and the other stages of creating projects. For planning purposes, the durations of the activities are studied *off* the diagram in routines suitable for repetitive calculations on a large scale using computers. In Fig. 2B, the lengths of lines representing each activity have been drawn in proportion to their typical durations, to indicate that the work content per activity tends to decrease as the number of simultaneous activities increases. These trends are shown in Fig. 2C.

The product of work content per activity multiplied by the number of simultaneous activities at each moment gives the rate of use of resources. This product follows a characteristic rise and fall of intensity of work in the evolution of a project as already shown in the upper part of Fig. 2A. The integration of this leads to the *S* curve or *Ogee* relationship between the accumulated total

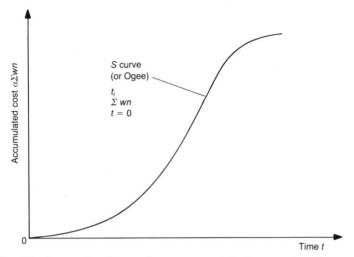

Fig. 2D. Accumulated cost of resources used during a project (S curve)

cost and the time that elapses in the evolution of a project. This is depicted in Fig. 2D.

Records of the expenditure during projects demonstrate this characteristic S-curve form, although the curvatures and skew vary, depending on the number of simultaneous activities and their durations.

Diagrams such as Fig. 2B give an indication that the complexity of activities changes during the evolution of a project. It is further complicated in most industries by division of the work for a project among specialist firms or departments within a firm. One example of this, found in nearly all industries, is the separation of the people employed to design from the consequent activities of manufacture and construction.

Stages of work

Stages of work as indicated in Fig. 2E are typical of engineering projects. Each stage marks a change in the nature, complexity and speed of activities and resources employed as a project proceeds.

The durations of the stages vary from project to project, and sometimes there is a delay between one stage and the next. The stages can also overlap. Fig. 2E shows the sequence of starting these stages. It is not meant to show that one must be completed

ENGINEERING PROJECTS

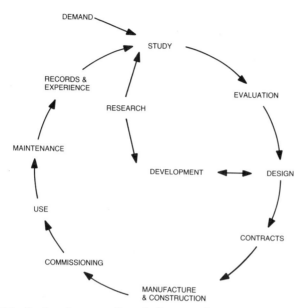

Fig. 2E. Cycle of stages of work for a project

before the next is started. The objective of the sequence should be to produce a useful result, so that the purpose of each stage should be to enable the next to proceed.

Initial stages

The cycle shown in Fig. 2E illustrates the STUDY of ideas for a new project drawing on the results from RESEARCH indicating new possibilities, RECORDS & EXPERIENCE from previous projects, and an estimate of the probable DEMAND for the goods or services the project might produce. These three sources of information, differing in nature, reliability and origin, ought to be brought together at this stage of a project.

The relative importance of information from research, demand and records & experience depends on the extent of novelty of the proposals and how far innovations as well as adaptation will be required in its design, but all three sources of information are relevant to some extent.

The cycle then proceeds to the EVALUATION of proposals, to compare predicted cost with predicted value. In emergencies, this stage is omitted, or if the project is urgent no time may be used in trying to optimize the proposal. More commonly, alternative proposals have to be evaluated in order to decide whether or not to proceed and how best to do so in order to achieve objectives. The results can only be probabilistic, as they are based on predictions. Their reliability varies according to the quality of the information used, the novelty of the proposals, and the amount and quality of the resources available to investigate the risks which could affect the project and its useful life.

Selection

Repetition of the work up to this point is often needed after the first evaluation, as its results may show that policy in meeting a demand needs to be reconsidered or made more precise, or because the conclusions of the evaluation are disappointing and a revised proposal which is more likely to meet a demand is needed. Expenditure is required to improve the reliability of the predictions. Repetition of the work may also be needed because the information used to predict the demand for the project has changed during this time. The study and evaluation stages may be repeated several times.

The STUDY and the EVALUATION stages are therefore closely related, and the two together are commonly known as the *feasibility study* for a project. They often proceed unsteadily because of uncertainties affecting decisions. The results are bound to remain uncertain, as calculations are matters of assessing probabilities in respect of the means of realizing the project, and probabilities that the result will meet a demand when it is ready to do so.

Estimates of the latter are particularly uncertain in considering a capital project which only indirectly meets public demand. Nevertheless a specification, budget and programme must normally be decided, albeit with contingent margins, as the decisions made at this stage should define all that is to follow.

The conclusion of this work may take time. Its result is quite specific: selection* or rejection of a proposed project. If the project is selected, the activities change from assessing *whether* or not it should proceed to deciding *how* best it should be realized and to specifying *what* needs to be done.

---

*Sometimes called 'sanctioning' a project.

Design, development and research

Design ideas are usually the start of the STUDY stage of possible projects, and the alternatives investigated before estimation of costs and evaluation of whether or not to proceed any further. The main DESIGN stage of deciding how to use materials to realize the project usually follows evaluation and selection, as indicated in Fig. 2E. The decisions made in design determine almost entirely the quality and cost and, therefore, the success of a project. Scale and specialization increase rapidly as it proceeds.

DEVELOPMENT in the cycle is the experimental and analytical work carried out to test the means of achieving a predicted performance. RESEARCH ascertains properties and potential performance. The two are distinct in their objectives. DESIGN and DEVELOPMENT share one objective, that of making ideas succeed. Their relationship is therefore important, as indicated in Fig. 2E. Research and development activities are often coupled in firms. They may share experimental and scientific resources, but organizationally it is DEVELOPMENT and DESIGN that need to be closely related for successful projects.

Most of the DESIGN and supporting DEVELOPMENT work for a project usually follows the decision to proceed. They may be taken in sub-stages so as to investigate novel problems and to review predictions of cost and value before a greater investment of resources is undertaken.

Manufacture and construction

Most companies and public bodies who promote new capital expenditure projects employ contractors and subcontractors from this stage on to supply equipment or carry out construction. For internal projects within firms there is the equivalent internal process of placing orders to authorize expenditure on labour and materials.

Although it may not be distinct in practice, the CONTRACTS/ORDERS stage has therefore been shown next in the cycle to denote the formal step of authorizing the stages that follow.

There follow the largest scale of activities and the variety of physical work in MANUFACTURING and CONSTRUCTION to realize the project.

Sections of a project can proceed at different speeds in design and consequent stages, but all must come together for COMMISSIONING and handing over for USE. The project cycle has then reached its final stage. It should then be meeting the specified objectives of the project.

The problems in meeting objectives vary from project to project. They vary in content and in the extent that experience can be adapted from previous projects in order to avoid novel problems. The criteria for evaluation also vary from industry to industry, but common to all projects is the need to achieve a sequence of decisions and activities as indicated in Fig. 2E.

Cancellation

A project might be cancelled at any stage, owing to changes in the demand or to problems which exceed predictions. The risk of cancellation cannot be avoided. The decision to start a project must be based on probabilities, and so some deviations from expected value and cost can occur on all projects.

These deviations are not necessarily worse than expected, but when they are worse this can be serious. Optimism in the selection of projects tends to be exposed as the work proceeds, but if all proposals were to be viewed pessimistically at the start, some projects of real worth might never be selected. The loss if this happens is less obvious than if projects that are no longer worth it are continued, but it is also a loss. Projects have to be selected despite uncertainty in meeting the demand economically, but the riskier ones can be taken in more stages, with checks between each stage for deciding whether or not to continue.

Information flow

The cycle shown in Fig. 2E depends on a clockwise flow of information. The main circuit represents the accumulation of engineering RECORDS & EXPERIENCE that can be used directly or adapted in decisions in respect of proceeding with a new project. The cycle is also open to new information: that from DEMAND indicating new objectives; that from RESEARCH indicating new means of achieving objectives. These different forms of information have to be equated in the decision of whether or not to proceed with a proposal. From then onwards there is an amplification into detail in the way indicated in Fig. 2B.

The chain of decisions and flow of information can be complicated and incomplete where the cycle is broken, because those who design and realize projects are usually specialists separated from

ENGINEERING PROJECTS

their customers who accumulate experience of the results. The resulting risks of divided interests and faulty communications are considered in Part II of this book.

Feedback of information

The ideal flow of information may also be complicated by secondary currents of information during work for a project, as shown in Fig. 2F. These consist of feedback from activities for the project, as distinct from the initial feed of records and experience from the previous projects indicated in Fig. 2E.

Figure 2F indicates that the emergence of new ideas during the design of a project can alter the estimates of market demand. The conclusions of the feasibility study have to be reviewed as a result. Errors in the predictions used can make it necessary to revise earlier decisions. As is considered further in the next chapter, the decisions in the selection and design of projects are made in the expectation that solutions of consequent detail will be found when the time comes to study these. During design there may be feedback of

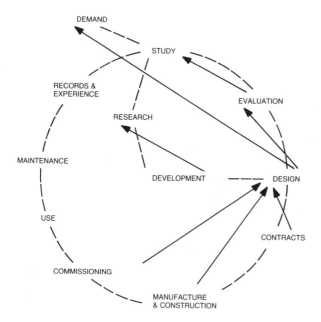

Fig. 2F. Feedback of information during work for a project

comments on drawings and specifications from manufacture and construction, some of this being comment requested on 'preliminary' drawings, etc., issued for the purpose, some being corrections, and some being later changes to follow decisions made in manufacture and construction.

Other feedback shown is from the contracts stage as committed costs become known. These costs are the test of the validity of the predictions used in the evaluation stage and of efficiency in keeping to objectives in the consequent work. Finally, during the commissioning of novel or complex equipment, the design calculations on the use of a project may have to be revised according to the results of measurements of performance.

The more novel the project and the importance in design of innovation relative to adaptation from previous projects, the more that feedback currents as shown in Fig. 2F are likely to affect the prior decisions made in the selection and in the realization of the project. The greater therefore is the need to plan ahead in order to anticipate problems and decide what flexibility and spare capacity to include for future changes.

Figure 2F indicates that DESIGN is the stage most complicated by feedback. As said earlier, it is the stage of decisions which most decide the quality and cost of a project. Attention to the relation of design to other stages in the cycle of work will therefore be our chief interest in the chapters that follow.

Interdependence of projects

The diagrams shown so far are models of what may be typical of the sequence of work for one project. Projects are rarely carried out in isolation from others, because of the potential advantages of sharing expertise and other resources. At the STUDY stage, alternative projects may be under consideration. In the EVALUATION stage, alternative projects may be competing for selection. Those selected are then likely to share design resources with others which may be otherwise unconnected, but will therefore be in competition with them for the use of these resources. This is the case through all the subsequent stages in the cycle.

A project is thus likely to be cross-linked with others at every stage shown in Fig. 2E. These links enable people and firms to specialize in a stage or sub-part of the work for many projects. Engineering resources should therefore be organized to try to achieve the potential value of concentrating on the needs of each project and the potential economy of people and firms specializing in a stage of work for many projects.

The consequence may be that any one project depends on the work of several departments or firms, each of which is likely to be engaged on a variety of projects for a variety of customers. In all of these organizations there may therefore be conflicts in the utilization of resources to meet the competing needs of a number of projects, and each promoter investing in a project may have problems in achieving the sequence of activities which best suits his interests.

Feedforward to further projects

Figure 2E indicates that the STUDY of ideas for a possible new project should draw on the RECORDS & EXPERIENCE accumulated from previous projects. For this purpose, the cycle for a project should be seen as one of a family, and each in form more a spiral than a closed cycle, so that every project starts from what has been learnt from previous ones and also contributes its lessons to those that follow.

Ideally, all the knowledge from all previous projects should be available to every engineer and should be applied to every new project. Human capacity so far limits the extent to which this can be achieved, even with computer aids that include expert systems. Experience alone is not enough, as it may be inaccurate, limited to memorable problems, and, at worst, give confidence only in repeating mistakes without the understanding of why they occurred. Experience is very valuable, but needs to be analysed to show what are the causes of successes and failures. Appendix C sets out principles and procedures for reviewing projects to learn the lessons and to produce usable recommendations.

Further reading

Stallworthy E. A. and Kharbanda O. P., *A guide to project implementation*, Institution of Chemical Engineers, Rugby, 1986.

Leech D. J. and Turner B. T., *Engineering design for profit*, Ellis Horwood, 1985.

Lock D., *Handbook of engineering management*, Heinemann, London, 1990.

3 Engineering decisions

Decisions in design

In chapter 2, the DESIGN stage of projects was seen as the vital process of engineering decisions which transform the predictions of the demand for products or services into instructions for a project to produce them. This process is now considered in more detail, as the basis for studying systems for the organization of this work.

Decision sequences

Observations by D. L. Marples of design under way in engineering demonstrated how it proceeds in a sequence of decisions in the pattern indicated in Fig. 3A. The decisions proceed in a branching sequence which begins with a few general decisions which affect all that follow and diverges thereafter into progressively more detailed choices.

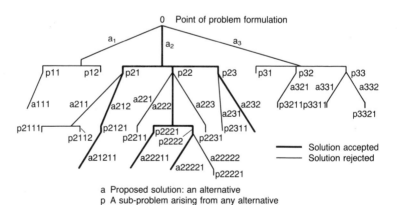

Fig. 3A. Decision trees in engineering design (D. L. Marples)

ENGINEERING DECISIONS

In this figure, the first decision in proceeding with the design of the project is shown as a choice between three alternatives labelled a_1, a_2 and a_3. The pattern shown is similar to Fig. 2B, but with the addition of these choices.

Each alternative solution is predicted as leading to subsidiary decisions. The choice of solution a_1 leads to decisions p11 and p12. The choice of the alternative a_2 leads to p21, etc. There may, in turn, be choices in deciding how to solve each of these. For some decisions there may not appear to be any alternatives, or not much may be known about the consequential decisions which would follow any of the possible solutions. Or there may be several solutions which require only a little adaptation of ones known from previous projects.

The number of decisions step by step and the number of alternative solutions can vary widely from project to project and within a project. Decisions in the process can also vary greatly in the extent that they are analysed and decided formally, depending on the number of people involved and their familiarity with the consequent decisions, but the common characteristic of projects is that a series of decisions leads to progressively more numerous smaller decisions.

An analysis of the sequence of decisions for a project should therefore indicate the links needed between firms, subcontractors or groups of people in a firm when sections of the process are divided among them. The pattern of decisions shown in Fig. 3A indicates that initially the vertical links between the people who have the knowledge and authority to make the decisions may be most important. The lateral links become increasingly important as more parallel decisions are taken simultaneously.

Transformation of information

The language alters in this process, as the dimensions of decisions change from those which measure customers' needs or supposed needs to those which state the results of design decisions on how to try to achieve those needs. This has been illustrated by D. Ramström and E. Rhenman, and their diagrams are reproduced in Fig. 3B.

As also observed by D. L. Marples, the initial decisions are stated in abstract terms such as the performance required. The information becomes more specific step by step, through estimates

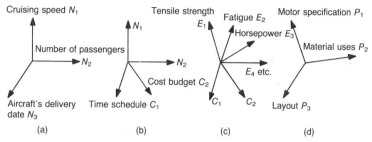

Fig. 3B. Transformation of dimensions from need to product: (a) the dimensions which measure customers' needs; (b) management's definition of the requirements; (c) the engineers' dimensions in problem-solving; (d) the ultimate solution (D. Ramström and E. Rhenman — Example of an aircraft project)

of costs, calculations of stress, etc., and ending with the final detail of shape, size, materials, finish, etc., giving the instructions for manufacturing and construction. Transformation of information is thus required in the cycle of work for a project shown in Fig. 2E, in chapter 2. Fig. 2E shows the ideal flow of information. To achieve this flow, the information produced at every stage must be in a form usable for the next stage.

Critical decisions

Figure 3A indicates the pattern of what have been called the critical decisions in design. Many other decisions have to be made, such as in planning the supporting development tests or in choosing methods of calculation. The critical decisions are those in which a choice is made between alternatives, and the subsequent work proceeds as if that decision were final. The nature of the work changes because attention turns to the consequent decisions.

Each decision in the sequence is a result of the prior decisions. Progress through the sequence therefore depends on the formulation of the initial need, and then choosing the solutions to the consequent problems after considering the probabilities that the consequent sub-decisions can in turn be solved. Mathematical analysis and development tests are increasingly the means of predicting the outcome of proposals. If a problem is a novel one, at least to the people involved, several possible solutions may have to be investigated in parallel, until the consequences of a choice between them can be predicted sufficiently for a decision to be made.

Simultaneous investigations of the consequences of several proposals consume more resources at the time, but the consideration of detail before commitment should make it possible to make the subsequent decisions more rapidly. Study of more than two solutions in parallel costs resources, but some reported observations indicate that investigation of the detail of at least two alternative proposals may be valuable in the feasibility studies stage of novel projects in order to see differences between them in ideas, in interpreting objectives and in the perception of consequent design problems.

Not all the consequences of design choices can be studied economically before critical decisions have to be made. A great expenditure of resources would be required to investigate all the alternative details before the decision is made to proceed with a project. Every decision therefore involves some risk. Success in completing the sequence of decisions depends on foreseeing that the consequential problems will be solved. Deliberate risks have to be accepted in limiting expenditure on the prior investigation of all consequences.

The greatest risks are the result of failing to foresee problems. The observations in firms suggested that personal experience is most likely to lead to the perception of problems that would follow an innovation. It does not follow that experience is the best basis for making a decision, only that it should be utilized to show what questions should be asked when any choice is considered. The decision-makers can play safe by making the best possible use of design solutions repeated or adapted from previous projects.

Innovation

From this we can conclude that the design of a project with novel objectives is likely to proceed successfully if innovation is avoided. This policy was observed to be successful in the studies by Marples. Novelty was *not* sought by the people making the critical decisions in the design of projects which had novel objectives. Only where known solutions cannot be repeated or adapted must new ideas be found. The need is a consequence of the initial decision of selecting the project. Whether envisaged or not, innovation may be necessary in order to avoid having to go back and change earlier decisions and perhaps to reverse the initial decision to start the project at all.

Observations in development work indicate that ideas for solving design problems are most likely to come through individuals in a group who have technological interests wider than their immediate needs and have the time to get involved in others' problems (T. J. Allen). These people are important for their intellectual variety, a quality that is independent of experience or specialized education. Experience and analytical abilities are essential to the assessment of the possible solutions to a problem; they are not reliable sources of new ideas. The perception and solution of novel problems before they become critical requires people with varied experience and some uncommitted resources.

The extent of division of work developed in large-scale manufacturing and bureaucracies should therefore not be assumed to be appropriate to design and development. The least division of work and loose linkings of people adaptable to the problems of each project are almost certainly more effective. That does not mean that their work should not be planned. The greater the uncertainty of decisions, the greater the value of looking ahead to foresee the flexibility and margins of space, etc., which may be worth providing for potentially critical decisions.

Development

Development may be required during design to test the proposed solutions to problems and so to raise confidence in the probability that proposals will perform as predicted.

In light engineering, such as electronics, development can proceed as part of design. Circuits can be tried, recalculated and altered rapidly. In heavier engineering, this close and flexible working between design and development is not possible because of the time required to make and test experimental models. Physical separation is often necessary in order to utilize suitable facilities. Attention to the links between design decisions and development work is then needed to achieve their common objective.

Classes of problems

From the observations in firms it was concluded that the difficult problems in the design of a project could be classified according to their tractability, the criteria for assessing them, and the resources required for their solution.

The most tractable class of problems is those that can be solved by further detailed study. The instances observed were problems which it was perceived could be solved given attention and some geometrical ingenuity in studying dimensions and the use of space. The design of energy systems and road networks, and other problems adapting from previous examples are similar. The effects of decisions in their design are largely predictable from experience. Design resources are required to proceed, but the sequence of decisions from the choice of project to the completion of detail should be completed satisfactorily if all decisions are of this class.

The problems are less tractable if the solutions available cannot be shown by analysis to satisfy engineering values such as safety and reliability. Development is needed to test these. Progress with the sequence of decisions may have to await the results.

Less tractable again are problems which it appears can be solved only by revising a prior decision. This may be necessary if no solution has been found to be satisfactory or if time or other resources are not available for development tests. The solution is achieved only by recycling the prior decisions.

The least tractable problems are those attributable to the natural properties of materials, etc. These set limits which can rarely be altered during the design of a project. The results of research may have provided new information on the properties of materials to be considered at the start in studying the feasibility of proposed projects. Application of this information during the design of a project can depend on advice from specialists. They may obtain further information during this time, as continued research is expected to demonstrate new and improved possibilities, but this is unpredictable and design decisions cannot depend on such possibilities turning up. As suggested in chapter 2, engineering projects should be selected using the evidence available at the start of design.

Clearly, the less tractable problems are best avoided if design of a project is to proceed. Choice of a solution of one design problem is tantamount to choosing to solve the consequent sub-decisions. Risks can be taken by not attempting to predict all these. The result may be that decisions have to be changed. Changes in demand and mistakes in decisions may also cause this. Only a limited extent of predicting consequences may therefore be worthwhile at the start of the design of a project, and the risk must be accepted that extra time and resources may have to be used if changes to data, or mistakes and intractable sub-decisions occur.

Decision processes

Many writers have specified procedures for making decisions in design. Although the evidence is incomplete, there is considerable agreement that every decision proceeds in several steps.

These steps can be part of the unconscious skills of an experienced person. Definition of them may then be too elaborate to be of value, especially when the detail of the design of a project is being decided. These deliberate procedures are more likely to be of value when the solution of a decision depends on collaboration between several people. The methodical analysis of choices is preferable in all critical decisions.

At least three steps can be identified as essential in a design decision, as shown in Fig. 3C. The first step consists of the definition of the need and the criteria for solving it. The second step is that of evolving solutions. The third step comprises the analysis and then the evaluation of these solutions. Formal procedures for this require that the analysis of objectives is kept distinct from the analyses of possible solutions. These impersonal methods should also require that estimates of intrinsic probabilities are kept distinct from the subjective probabilities given to them by the people taking part. If a solution satisfies the requirements, the resulting decision leads to the consequent decisions. If no solution satisfies the requirements or the requirements then change, the earlier information must be reconsidered and the cycle repeated.

The step of synthesis to evolve solutions depends on information to adapt and the conditions to innovate. The step of analysis to test these solutions depends on experience to perceive problems and the engineering science to analyse them. But first the requirements and the criteria for satisfying them must be defined. They govern the decision, and it is the requirements and criteria which have to be reconsidered if no satisfactory solution is found.

If such a solution cannot be found, the choice is between taking more time to search for a new solution or to go back and reconsider the prior decision. It is safer to decide the latter. On the other hand, it may be natural for the people working on a problem to be optimistic and to expect that their ideas will turn out to be satisfactory. Caution and optimism in these ways can be in conflict,

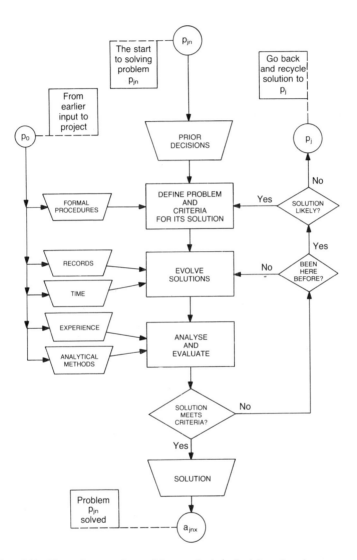

Fig. 3C. Formal steps in making a design decision (p_{jn} is any one decision in sequence shown in Fig. 7, and a_{jnx} is chosen solution of that decision)

particularly after some time has been used in considering novel solutions to a problem. Such conflicts may best be solved by following a deliberate procedure to define the requirements, list the possible solutions and state the criteria for analysing them.

Organizational consequences

In this chapter we have considered the evidence that the design of a project depends on a sequence of decisions, in order to be able to use this to see the links needed between firms and groups of people in a firm.

Most important is the formulation of the initial statement of requirements. The rest follows from this first critical decision. Subsequent decisions proceed more in parallel. Some decisions may be uncertain and complex. Others may be divided into independent decisions, and so on, to the details that form the link between design and the consequent stages of work for a project. Each decision may require experience, analysis and time to consider solutions with different varieties of specialist and development facilities according to the class of sub-decisions. The system of organization should be designed accordingly. It should link all who are involved, from the initial definition of requirements and criteria to the development and completion of detail

Further reading

Allen T. J., *Managing the flow of technology*, MIT Press, Boston, 1984, 2nd edn.

French M., *Conceptual design for engineers*, Design Council, London, 1985, 2nd edn.

PART II

Relationships between organizations

4 Manufacturing

Change and complexity
The development of manufacturing by specialist processes in factories is relatively recent in the evolution of civilization. It was greatly accelerated by the industrial revolution that began 200 years ago. As a result, society has become more complex and is increasingly dependent on investment in technical skills, the construction of the manufacturing and distribution systems, and the management of all these. The developing countries are tending to follow the same process of replacing local crafts and self-sufficiency by dependence on factory-produced goods. The attraction is economic, as industrialization promises higher standards of living.

Manufacturing has tended to become continuous, as now seen in the newer industries such as chemical manufacturing, and to be concentrated in larger systems so as to gain economy of scale by sharing capacity and skills. Changes in demand, social values and competition counter this. A steady state has a limited life and because of continuing innovation is generally shortest in the newer industries. The result is that the products of most factories vary in type, size, complexity, relative importance, timing and length of life in production. Concentration and continuity of manufacturing is also limited by external restraints to these tendencies, such as scattered, seasonal or varying inputs, or where output must be dispersed to provide a local service. Food processing includes some obvious examples of these restraints. Manufacturing industries thus tend towards greater concentration, but differ in the extent of their concentration of production.

This trend continues in all countries. It is not complete and may never be, because there is also a continuing demand in industrial countries for handmade goods and personal services that are distinct from factory products. There also remain traditions of industrial training and employment which relate more to old crafts than to future prospects. These exceptions indicate that society and individuals are influenced by more than economics, but after necessities have been supplied. We may also observe that the quality of factory products has become more attractive, and that many of today's 'necessities' were considered luxuries generations ago. The use of more goods and services soon becomes habitual. Greater happiness may not follow, but the process could be reversed only painfully. In general, production is now increasingly dependent on prior capital investment in systems and products of greater complexity but shorter economic life. People and communities have therefore become more and more dependent on engineering systems for providing goods and services.

Specialization of firms

Many developments of machines in the industrial revolution were begun by the users, such as the owners, engineers and managers of mills and mines who looked for more powerful, reliable and cheaper means of production. When successful, their new machines and expertise could also be sold to other firms. Some firms then divided their activities, one part continuing the original production and another developing the new business of making machines that could be sold to other users. The latter became the main business in what were then called 'engineering' firms.* Thus emerged the separate firms of machine tool manufacturers selling to all other parts of industry. New engineering firms were also formed by people with special experience or to develop the ideas increasingly resulting from investment in research.

The pattern of the evolution of a new industry can be illustrated by many examples such as the history of the development of steam engines, textile machinery, shipbuilding and electrical power systems. New firms emerged to form each of these industries, at first often designing to suit each customer, and making much or all of the product as special for each order. Many firms set out to try to meet a newly growing demand. Initially, they might prosper as innovators, but as the demand became more established they had increasingly to compete by reducing costs or by offering

---

*Economists use the word 'engineering' to mean manufacturing.

customers improved products and more services such as installation, maintenance, etc. Competing firms then tended to join together to pool resources to innovate, share risks, reduce competition and gain economy of scale. Further new divisions and new firms also continued to be formed to exploit new technologies, a recent example being the growth of the electronics industry and its emergence as a distinct new branch of electrical engineering. Such a new industry can take over markets formerly supplied by others, but is itself increasingly likely to reach an affluent peak followed by concentration on products and services for particular markets.

The evolution of engineering

The agents of industrial revolution were the engineers who added analytical methods to established craft knowledge. They evolved the practice of systematic design in developing machines and materials. Also, they shared their theories and experience, in meetings of professional societies. Innovation became deliberate. The 'greatest invention of the nineteenth century was the invention of the method of invention' (S. G. Checkland). It no longer depended on individual genius and discovery. Trial and error was augmented progressively by the growing engineering science that could be used to predict mathematically the probable performance and risks of proposed machines and systems.

Errors have not thereby been eliminated. The economic risks have increased. As machines became more complex and part of systems, their success could be affected by relatively small mistakes or by uncertainties in scaling up from an experiment to working conditions. Nor has innovation been universally welcomed, by engineers or by many of those who could be affected. To innovate successfully, individuals and firms have needed greater discipline to agree objectives, analyse risks and to plan projects to try to anticipate problems.

Advances in machines and materials go together, new types and qualities of materials being essential to making machines of greater power, size and accuracy which, in turn, are used to produce other machines and materials and so on in the sequence, ending with the systems of machines used for making and delivering products to customers. The evolution of machines well into the industrial revolution remained dependent on the strength and reliability of

PRINCIPLES OF ENGINEERING ORGANIZATION

iron, wood and stone. Materials were a limitation on the use of power until the development of steels, electrical materials and plastics. New materials have made possible the design of machines of increasingly greater power, speed, accuracy and reliability. In this there has been increasing specialization in producing the machines and the materials.

Interdependence of firms

An obvious result in Britain and other industrial countries is the increasing interdependence of firms in the manufacturing of materials, machines and, ultimately, of consumer goods and services. Nearly all the foods, clothing, medicines and other necessities of life are now manufactured by processes which use the products of other firms. Complex patterns of manufacturing and assembly of mechanical, electrical, structural and other materials and components lead eventually to the final output of consumer goods and services. A simplified picture of this is shown in Fig. 4A.

Any firm or part of an industry thus relies on the products and

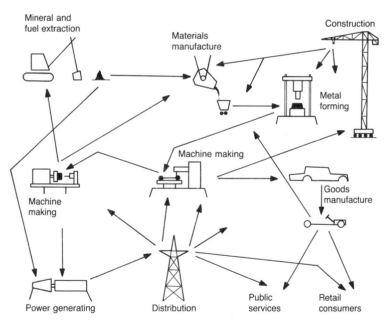

Fig. 4A. Sketch illustrating interdependencies between industries

services of firms in other sectors of industry. For a project such as a new factory or a renovation or improvement, a firm expects to be able to buy the machines, structures and services it needs by contract from other firms which specialize in meeting such demands. In most industrial countries, suppliers are competing for these contracts. The customers then have a choice in a system of transitory relationships which can be flexible. This is valuable as skills and resources can be brought together as needed for a project, but requiring both parties to establish relationships between them for each contract.

Most firms in the capital sector are specialists supplying only one or some of the types of machine and services that may be required for a new manufacturing system. Many may have to be employed on one project. Several of these firms may be owned by one parent company, as a result of subdivision, acquisition or merger, but mostly they operate as separate firms. A new manufacturing system therefore tends to depend on firms employed in parallel to supply machines, etc., which in use are operated in series.

The suppliers in turn employ other firms as subcontractors to supply components, materials and services, with consequently a greater separation between suppliers and users.

Figure 4A illustrates this separation between suppliers and users. It occurs in nearly all manufacturing, although the extent of the division of work varies from project to project and changes with the rise and fall of firms.

Projects

A new product may depend on a change of manufacturing system, but not necessarily. The design or improvement of a system may consist of rearranging existing machines, systems, controls, services and buildings. It can include innovation in any of these, and may utilize a variety of novel and conventional machines, etc., from several firms.

Engineering for manufacturing is therefore employed on two categories of project

- new or improved *products*, particularly their design and development for large-scale manufacturing
- new or improved manufacturing *systems*, particularly the design of the machines, factories and services.

The above distinction between the two categories of projects is important, as a new product has the objective of meeting an external demand whereas a new process is only an indirect contribution to meeting it.

The relationships between firms for a new or improved system are considered in the rest of this chapter. The relationships within the customer and supplier firms are considered in chapter 8.

Project relationships

Figure 4B illustrates the relationship between a supplier* and the user of the machines for a manufacturing systems project.

The cycle of project stages that was discussed in chapter 2 is the basis of this diagram. One area of shading on it indicates the activities of the contractor supplying a machine or system, labelled as in the *capital sector*, as its product is an asset purchased by the promoter. The promoter is labelled as in the *consumer sector*, as it uses the machine. The radial width of the shading indicates the changing intensity of their engineering activities during the cycle. The relationships are similar between the promoters and the contractors who supply supporting services.

Contracts

Figure 4B indicates that the contractor usually designs and develops the machines and systems for a new manufacturing system, and on receiving a contract from a promoter becomes responsible for supplying, installing and at least partly commissioning them. The promoter usually decides the performance required from the completed system, on the basis of his engineering experience or the advice from consultants, but often also drawing on contractors' proposals and licensed processes.

There can be several contracts in parallel, with the promoter and contractors each supplying their specialist equipment or services for a project. Or there can be one comprehensive contract

---

*A supplier is a contractor usually responsible for designing, supplying and often also for installing and testing a machine or system. 'Contractor' is therefore used in the rest of this chapter.

The word 'promoter' is used to mean the firm which is purchasing and is the future user of a manufacturing machine or system.

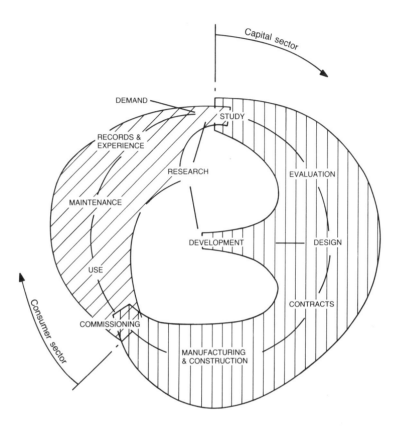

Fig. 4B. Project roles of supplier and user of a manufacturing system

and the one contractor is responsible for obtaining all that is needed.

A firm that proposes a new manufacturing project has a choice, therefore, of whether to buy the requisite machines and systems from each of the specialist firms who can design and supply them, or whether to invite one firm to provide the complete project.

The choice varies from project to project, depending on a promoter's experience and his priorities between quality, economy, speed and flexibility. The obvious principle should be to choose contract arrangements which are likely to meet all parties' objectives and therefore to avoid conflict. All experience shows that to achieve this demands attention to the choice of number and type of main contracts and subcontracts, and to the style of relationship between all parties from start to finish.

The simplest type of contract is one in which a contractor undertakes a fully defined package of work for one promoter, and is paid a fixed amount for doing so. Contracts for purchasing standard pieces of equipment, materials and routine services can be as simple as that. Contracts for systems are usually more complex, because both parties share the design and other risks.

Separating responsibilities for stages and types of work may be logical if design requirements are uncertain at the start, if the promoters project team is newly formed, or if contractors have limited capabilities or knowhow relative to the size or type of work needed for a project.

The potential advantages and disadvantages of these choices are summarized in Appendix B. As may seem obvious in principle, although not always achieved in practice, the success of any contract depends on the ability of the promoter to define his requirements and to assess his risks, and on the capacity, motivation and supervision of the contractor in planning and controlling all his responsibilities.

Innovation

The characteristic of manufacturing illustrated by Figs 4A and 4B is that firms design and make part or all of a product which is used in turn by others. A new product depends on the means of manufacture being available, but the makers of the machinery it needs do not gain the direct experience of its use by others. In these common circumstances, the completion of the cycle of information from project to project depends on continued links between supplier and user during the operation of the machines. The principal users are best placed to take action to make these links effective, as their specialist engineers in the planning and maintenance of manufacturing systems can accumulate the experience and records that should be analysed in order to decide how to improve the study and design stages of new projects.

The relationship necessary between designer and user to solve faults are not likely to be sufficient to achieve this feedback from experience and records. The faults that are reported may not be typical, and the designer may not know about changes made by the user. A survey of performance and problems is needed as a routine activity to review all projects after completion. For this, a set of users can act collectively to supply records to be analysed by a research institution or a similar experienced body that would feed the results back to the users and the designers.

A disadvantage of an established division of interests is that innovation in manufacturing tends to be dependent on prior investments by others who lack the direct incentive. The typical pattern is that innovation in the consumer sector may depend on the machines available, and that these in turn may have to await the results of investments by other firms in developing improved or new machine tools, materials, control systems or supporting services. New production for which there is a demand may therefore be delayed unless these dependent capital investments have already been stimulated. The opportunity to innovate ahead of demand by developing an idea or special experience may have usefully stimulated the formation of specialist firms, investment in innovation being their initial action in order to develop a new product and so attract a demand for it. After that, this incentive may decrease.

The advantages of the division of work between manufacturing firms are in the division of risks, the concentration of specialist knowledge on specific objectives, and the flexibility in use of the products. For individuals and firms there are incentives in the chosen specialization. For their customers there is the advantage of being able to make use of successes.

Relationships between firms

From the general survey in this chapter, it can be seen that firms can have many relationships with others, some in a regular sequence of work divided vertically and others in occasional links for particular projects. The organization of a firm in manufacturing may therefore have to be flexible to relate to a variable mix of dependence on customers, contractors and sources of ideas and information, and all these relationships are subject to pressures for change to meet evolutions in markets, materials and technology.

As a promoter with projects which differ in their priorities, complexity, urgency and novelty, a firm may need an internal system of organization to employ suppliers differently for each project. As a supplier, a firm may need an internal system of organization to work with different promoters on different contract commitments.

Further reading

Diebold J., *The innovators*, Dutton, 1990.

Holt K., *Product innovation management*, Longmans, Leatherhead, 1988, 3rd edn.

Johne A. and Snelson P., *Successful product development — lessons from American and British firms*, Blackwell, Oxford, 1990.

5 Construction

Construction conditions
The obvious characteristic of construction is that projects are designed for the conditions of the proposed site.

Factory manufactured materials, components and construction plant are used, to minimize the costs and duration of construction, but much of civil engineering consists of designing and planning construction to suit the ground, weather and other conditions of each project. Production is therefore mainly a temporary process on a site. Design and construction planning are subject to uncertainties of the site conditions and climate, so both may have to be revised during construction. For most projects, it is economic to design what can be constructed by employing the skills, materials and plant which are commonly available, as these can be used flexibly.

Another difference between construction and manufacturing is that much investment in construction is to provide services rather than products, and in most countries is financed largely through central or local government.

Specialization of firms
In Britain it is traditional for one 'main' or 'general' contractor to be employed to construct a building or civil engineering project. Design is the responsibility of engineers and architects employed by the promoter.* As in all engineering, the main contractors employ subcontractors to do specialist work on site, provide local services and supply materials and sub-systems.

---

*In some civil engineering contracts in Britain and in related international ones the promoter is called 'the Employer'.

Comprehensive contracts are also used, perhaps increasingly so, with the one contractor responsible for complete design and construction. For some projects the alternative of several separate specialist 'trade' contractors are employed in parallel, with complete design from the promoter. The practice in other countries also varies between these alternatives. It is the promoters who choose, within any limits set by law or financing bodies. Contractors and others therefore have to be organized to operate in these different contract arrangements.

Traditional contract relationships in Britain

Separation of design and construction evolved in civil engineering in Britain in the great age of road, canal and railway building which began in the eighteenth century. The services of civil engineers able to design novel projects successfully were suddenly in great demand, especially to advise promoters on the cost and the planning of large scale investments.

There emerged the separate firms of public works contractors who became specialists in mobilizing people and the plant for the temporary demands of projects. The design engineers became consultants independent of the financing of construction, and mostly in partnerships employing qualified and more specialized engineers. A similar separation of roles had evolved between the architects and the building contractors. Architects and consulting engineers established professional firms, to be consulted by promoters to study ideas for new projects and to report on costs, problems and how to proceed. If a promoter decided to proceed, the consultant was then usually appointed to undertake design, prepare specifications and other contract documents, and to advise on choosing the contractor.

Promoter, engineer and contractor

In what has been called the 'traditional' procedure in Britain, the promoter also appoints the consultant to be responsible for the administration of the construction contract. In the contract, the consultant is named as 'the Engineer',* with powers and duties to supervise the contractor and to make decisions on design and

---

*Or various other titles in building and civil engineering contracts such as the Architect, Promoter's Representative or Supervising Officer.

CONSTRUCTION

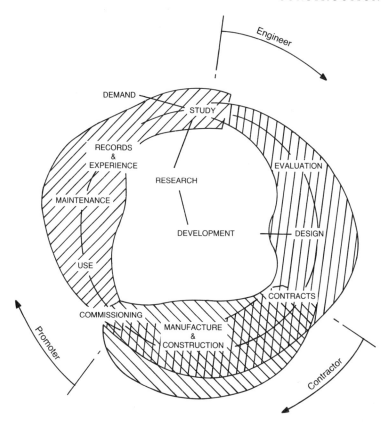

Fig. 5A. Traditional responsibilities in cycle of activities for a British civil engineering project

construction. In this system, the Engineer (with assistants) drafts and administers the link between the promoter and the contractor. The employment of a consultant in the traditional UK system to design and supervise the construction of a project is a different and larger role than that of a consultant employed to advise on particular problems.

Figure 5A illustrates the responsibilities of the promoter, the Engineer and the contractor in the cycle of activities for a project. The scope of work shared between these three is indicated by bands of shading. The changes in the intensity of their involvement in the engineering decisions as the work proceeds are shown

PRINCIPLES OF ENGINEERING ORGANIZATION

diagrammatically by the changes in the radial width of these bands of shading.

The promoter and the Engineer are shown sharing the feasibility stages of study and evaluation of proposals. The division of this work between them is likely to vary according to their relative experience of the type of project.

Figure 5A indicates that most of the design of a project is usually completed before selection of the contractor. This division of work can also vary in practice, and the designer may arrange to discuss construction methods with one or more contractors before inviting tenders. Contractors can also be permitted and encouraged to offer an alternative tender based on their own design for the project.

Under the terms of many British contracts, the Engineer is the channel of communication with the contractor on design and construction, as illustrated in Fig. 5B. The Engineer has powers to vary the type or quantity of contracted work, and to state the payments the promoter must make to the contractor for progress, extras and delays.

Specialization and continuity

In the construction procedure just described, the consultants and the contractors can develop their distinct specializations to offer

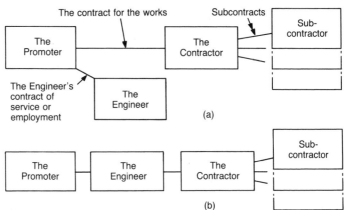

Fig. 5B. Contractual authority and formal communications links in conventional British system for civil engineering projects: (a) traditional contractual relationships between promotor, engineer and contractor; (b) communications channels in traditional system

to promoters. Their experience of design, site problems, construction and management thus becomes available to promoters who are engaged in a diverse programme of investments. As the demand for most construction work is not accurately predictable, the flexibility of this system of employing the specialist services of consultants and of contractors when required can be more economic and convenient to promoters than the establishment of their own engineering departments and construction resources.

A potential advantage of the traditional British procedure is that the Engineer provides not only engineering services and contractual advice to the promoter but in the contract between promoter and contractor is usually empowered to make decisions on construction problems and order variations. This procedure developed as a means of controlling payment for construction based on competitive tendering. Its disadvantages as a system are inherent in the vertical division of the sequence of work for a project, as summarized in Appendix B. Discontinuity is the price of employing the consultant and then the contractor, as information must be transferred and transformed from firm to firm, and each one in the series can learn only indirectly of the consequences of their decisions. Feedback of experience and improvements in methods therefore requires action outside the needs of each project.

Another discontinuity can occur when the main contract is awarded. Before this, sufficient detail of design must have been decided to complete the contract drawings and other documents to form a commitment of the contractor to the promoter. Methods of construction must therefore be assumed before the contractor is selected and his methods are known. Previous discussion with contractors does not ensure that they can agree innovations in methods that will be best to propose when the time comes to tender. When the contractor is selected, changes to design to take advantage of faster or cheaper methods of construction can therefore only be introduced at the risk of causing delay when construction should be accelerating.

These discontinuities are accepted in the system as followed for many projects. The discontinuity in applying promoters' previous experience is probably less important in many cases than the possible gain from drawing on experience of the design of projects for other promoters. The competence of the designers and the contractor are therefore important. As each in turn is responsible in a sequence, the weakest organization of the three can govern the quality of the result.

Although the Engineer's role and powers in contracts are not distinct in most other countries, similar problems occur in all systems which attempt to achieve the advantages of specialization despite the uncertainties of demand, changes, delays and the other risks of construction.

Authority−contractor relationships

The contractual links are simpler if the promoting body directly employs its own permanent engineering and other specialists rather than delegating design and other decisions to consultants. This has been common in public services in Britain. An organization which in this way combines the scope of promoter and engineer is here referred to as an *Authority*.

Within an Authority the design and contractual control of a new project can lead on from studies of demand, and they can be coordinated with the use and maintenance of existing structures and equipment. Although most authorities employ contractors to construct their projects, authorities working in this way can more closely link design and construction with their other responsibilities as promoters and users of projects.

Operating as an Authority is therefore attractive to local government and statutory bodies who have public responsibilities for the safety and convenience of services. The logic of the system for coordinating new construction work close to structures and equipment in use is also attractive to manufacturing and similar firms for projects to extend or alter part of an operating site.

More direct learning from experience and simpler communications during the work for a project are thus possible if promoting and engineering decisions are unified in this system. To achieve these advantages, an Authority needs to have a programme of sufficiently regular investment in projects to employ enough experienced engineers, managers and assistants without their numbers and variety becoming the basis of divisions of interests in the Authority.

The Authority−contractor system remains distinct from the traditional procedure if an Authority employs its own engineers but adds to them the services of a firm of consultants without appointing the consultant as the Engineer independently responsible for contract administration. This use of the services of consultants enables an Authority to retain direct control of a project but draws on extra and specialist manpower. It brings in the experience gained by the consultant on other promoters' projects, and it gives the consultant experience in return.

An alternative system to share experience and specialist manpower is for a group of authorities with related projects to form a joint engineering unit. This could be a particularly suitable arrangement for a series of related projects which are large and complex compared with the usual investments of each Authority.

The advantages of the simple Authority–contractor system are attractive if the technical problems of projects are not the same as those encountered by other promoters. The disadvantages are that the Authority's managers or engineers may not perceive that other experience is relevant to their work. Isolation may also make them uncertain in their standards. Ways of mixing experience have been discussed which can avoid these effects of isolation.

The outstanding disadvantage compared with the conventional system is that the engineering managers are only part of an Authority's organization and may not be effective in making engineering or contractual decisions which are contrary to opinions of other parts of the Authority.

The Engineer in an Authority

In this chapter, systems with a promoter directly employing permanent engineering manpower have been contrasted with the conventional system of delegating powers to consultants.

In practice, these systems can be combined. The promoter can decide to prepare invitations to contractors to tender for work under the terms of contract established for the traditional British procedure and can name his own employee as the Engineer. If so, the promoter has to accept that the Engineer has to administer the contract in accordance with the terms of that contract.

Innovation

Systems based on a division of work between contractors, consultants and their employers promoting the projects have the advantages reviewed earlier, but the coordination of their operations for a project depends on contractual communications which link the three.

The common training and professional associations of most of the engineers employed in all parts of construction assist informally in establishing communications on a project, but the temporary need and the division of interests between organizations make it difficult to innovate and to transfer experience to new applications. Some transfer is achieved when individuals move from firm to firm to take a new job. This is probably particularly valuable when the move is from one part of the industry to another, but studies of analogous situations in other industries have indicated the advantages of continuous mixing of experience and of direct relations between all concerned to achieve innovation. Although there is no definite evidence from comparisons, this simpler contract system should therefore be more effective in achieving improvements in practice.

Industrial projects

For the construction of industrial projects it is usual for contractors to be responsible for much of the design, starting from a performance requirement or specification issued by/or discussed and agreed with the promoter.

The building and civil engineering construction may be only part of a project, and be subject to changes made to suit the project as a whole. In examples of heavy industrial projects such as power stations and refineries, the sequence of activities ideal for all or some of the construction may be dislocated to suit the design, supply and installation of mechanical, chemical or electrical systems, machines and services. In examples such as large buildings, the installation of mechanical and electrical services can be the most critical phase on site.

Direct labour

Some promoters also directly employ their own permanent construction manpower and management. This has the advantage that they can work on maintenance or on the construction of new projects as needs change. As with engineering manpower, direct employment of any resources is logical if there is sufficient regular demand.

Public authorities in the UK are now usually allowed to employ their direct labour departments only by contract in competition with contractors.

Comprehensive (package-deal, all-in, turnkey) contracts

The alternative simplification compared with the conventional procedure in construction is for one firm to be designer and contractor.

As discussed in Appendix B, this system has the advantage that unity of the design of a project with the planning of its construction should save time, cost and disputes, particularly if the contracting firm is big enough to include the expertise in architectural design and layout of equipment and services for projects such as factories and other large buildings without the creation of internal divisions of interests between all these.

In proceeding to such a system the promoter can invite the interested contractors to undertake feasibility studies in competition, and can employ a consultant for advice in deciding the specification for inviting tenders. To benefit from this system, a promoter must complete his specification of needs before inviting tenders, and then leave the selected contractor with the responsibility for meeting the specification to his tendered price and programme.

A comprehensive contract is logical where unity of design and construction are important but where site uncertainties are relatively small. The system is therefore used for commercial and other projects where saving time and avoiding disputes are more important to the promoter than the detailed control of changes or of cost. It is also logical for it to be used on very novel projects to obtain unity of innovations in design and methods of construction.

If a promoter wishes to take advantage of this system for a project with foundation uncertainties, one possible way of proceeding is to place two contracts, one for the substructures under the flexible traditional procedure and then one for the complete design and construction of the superstructure. This would be a logical arrangement where there is variety and uncertainty in ground conditions but a large amount of conventional superstructure work has to follow.

Evolution of systems

Irregularity in the demands of each promoter does appear to have been a cause and perhaps the chief cause of the evolution of separate firms of consultants and of contractors. This system has been a means of distributing their resources flexibly.

In the evolution of construction the organizational problems within consulting engineers and of contractors have become less distinct. The consulting engineer has become increasingly the leader of large groups of professional engineers and assistants. He* has also had to make more contractual decisions on construction problems and methods, as a result of the changes in the terms of contracts that have tended to limit the contractors' risks. In this time the contractors have become more and more the employers of engineering expertise to plan the economic use of increasingly specialized plant and more complex subcontracts. Their management of people has had to be less authoritarian and instead the managers become more professional leaders and engineers. As a result, consulting engineers and contractors have more problems in common and should be more able to agree on unusual systems of organization for projects with unusual demands.

The result for most firms in the industry is that either they take the risk of greater specialization and so become dependent on particular demands, or they take the risk of growing in order to undertake projects of all sorts and sizes and so have to work simultaneously in many increasingly different ways to suit customers and collaborators. In order to succeed, the systems of organization within firms have to adapt to these changing and variable demands.

Further reading

Cox P. A. (ed.), *Civil engineering project procedure in the EC*, Thomas Telford, London, 1991.

Austen A. D. and Neale R. H., *Managing construction projects — a guide to processes and procedures*, International Labour Office and Heinemann, 1984.

Engineering construction risks — a guide to project risk analysis and risk management, Thomas Telford, London, 1992.

---

*Throughout this book the word 'he' refers to the role and means the man or woman in that role.

6 Joint ventures and consortia

Joint projects
Joint ventures and consortia are used increasingly in many industries and countries for temporary or selective cooperation to share resources and risks, and to enter new markets. Most typically they are temporary arrangements for the purpose of carrying out one joint project. Some are continuing arrangements for joint development work, a series of related projects or for the operation of completed facilities.

Definitions
Joint venture: Collaboration between two or more firms, usually agreed between them for a specific project.
Consortium: Collaboration between two or more firms through a jointly-owned subsidiary company, usually for a project or series of projects for a specific market.

The abbreviation JVC is used in the rest of this chapter when remarks apply to both joint ventures and consortia.

The aim of a JVC is partnership. JVCs are selective in that the partner firms continue with other business activities independently, sometimes in competition. They are systems for limited cooperation as distinct from amalgamation to form one firm, although working together in a JVC may be used as a test-bed to try out the potential benefits of continuing cooperation and may lead to amalgamation.

Horizontal collaboration
The least interdependence and simplest relationships between partners can be achieved if their activities can proceed in parallel and simultaneously, that is they are related *horizontally*.

Instances of this are

- manufacturers who share a common operation in their lines of production, such as oil companies jointly owning an offshore platform, chemical manufacturers sharing a factory for processing raw materials, and electrical manufacturers sharing a plant for making standard components
- promoters such as statutory authorities who have jointly agreed on designs of projects such as hospital buildings, in order to gain economy of scale by ordering a series of standardized structures
- consulting engineers acting jointly with economic consultants and local experts to advise governments of developing countries in their schemes of national and regional planning
- contractors who form a joint venture to promote a project, particularly for infrastructure projects
- civil engineering contractors who form a joint venture to construct a project but can divide the construction site into their own separate areas.

In many cases such a joint venture is also *homogeneous*, meaning that the partners are in the same sector of industry.

Vertical collaboration

More complex are *vertically* related activities, i.e. when partners depend on each other's activities in a sequence. Instances of this are

- a potential user and a supplier who collaborate in a joint investment, e.g. a manufacturer who partially sponsors the development of a new material by a supplier, or a promoter who invites contractors to share the cost of building a new factory and be paid from the operating income
- a promoter who forms a joint team with a contractor to plan and manage the design and construction of a project
- a joint venture formed by a civil engineering contractor who constructs a building, and a mechanical, electrical or process plant manufacturer who follows by installing equipment and services in it.

In these instances, the JVC is *heterogeneous*, meaning that the partners are in different sectors of industry or undertake different classes of work.

In these arrangements, the user and the supplier form a temporary vertical link that can overcome the discontinuities of experience and of interests which have evolved in all industries, as described in chapters 4 and 5. A vertical JVC makes it possible for the user's experience and records to be applied to a new project. By this means, the future user of a project can be a party to decisions.

The partners' expertise and interests are not the same. Their mutual interest, which makes the collaboration attractive at the start, can change as the investment proceeds.

Complex collaboration

Some JVCs combine horizontal and vertical relationships, as in the following instances

- process, mechanical, electrical and civil engineering contractors who jointly design and construct projects
- contractors who jointly contract to design and construct a multi-project complex made up of interdependent plants constructed by the different partners but sharing some subcontractors
- contractors, consultants, bankers and others who form a consortium to offer the design, financing and execution of a complete project, as has been the demand of some customers and governments in other countries
- joint working of European and local contractors to construct a large project in a developing country. The arrangement is not homogeneous because the firms differ in their expertise and the local contractors will probably have longer term prospects of work in their country which are not open to their foreign partner. This can therefore be a horizontal but heterogeneous JVC.

The partners' activities for complex projects can be related *alternately* (the first partner in a sequence of activities has further activities to carry out after activities by another) and *reciprocally* (partners depend on some activities by each other).

In all JVCs, the collaboration is only a part of the interests of the partner firms, and they otherwise continue to work in their established markets. Collaboration is likely to be of mutual interest only for a project that has abnormal uncertainties in demand or in design. If this is so, the firms should expect more recycling of decisions than is usual in their work, to adjust to feedback of information from partners.

Complexity and control

Promoters and financing bodies normally require the performance of a JVC to be guaranteed by its partner firms. Given this basis, a horizontal collaboration should be simple to operate as a joint venture if the partners have similar expertise and interests and can divide the work into independent packages. Parallel working of their normal systems of organization can then be effective in achieving sufficient linking of decisions during the project.

To achieve consistency, a single channel of communication with a customer is always better for all parties, as indicated in Fig. 6A, but control of the firms' operations need not be centralized unless some resource such as a common subcontractor has to be shared.

Vertical or more complicated collaboration may require formal linking and overlapping of the partners' systems of organization, not necessarily on a large scale, but starting before commitment to a project so that joint decisions are possible in a way unusual in their normal work. If all partners are used to this, a joint venture can be adequate.

If not, either a consortium will be needed, or the JVC will be replaced by one firm taking over and the others working for it by contract. If there is a conflict between a partner's commitment to the JVC and to its normal customers, the JVC may get no more attention than any other customer. Indeed, the JVC may get less attention as it is only a temporary interest. A joint venture is no

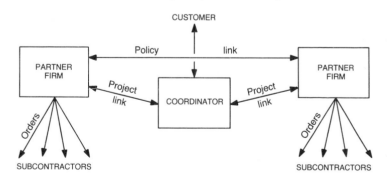

Fig. 6A. The joint coordinating role

more unified than a part-time partnership, particularly dependent on collaboration. A consortium is a means of committing its partners' resources to a joint project and of establishing a separate management of them.

A consortium with formal delegation of authority to a separate dedicated management may be needed particularly for a JVC based on a combination of commercial companies and public authorities, to avoid their differences in corporate autonomy and accountability causing operational differences in managing risks.

Figures 6A and 6B illustrate the choices for a system of control. Fig. 6A depicts a person or team with the role of coordinating the partners, a role which depends on influencing their managers who control the resources needed for a joint project. Fig. 6B shows that the partners have established a joint subsidiary which formally employs the partners by contract to carry out their work for a joint project.

A joint subsidiary firm established as shown in Fig. 6B can be strengthened if it is given the authority to commit its sponsoring firms and to accept risks in orders from customers. Logically, it then becomes responsible for profit and for the control of financing to cover mistakes and other contingencies. Accordingly, it must also be free to act as a main contractor and to employ its sponsoring firms in competition with others.

Vague arrangements falling between these systems are possible

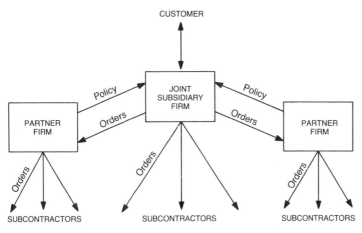

Fig. 6B. The joint controlling role

in joint activities for complex projects. These can be effective if a project is free of mistakes or changes in demand or if mutual longer term interests lead to the joint sharing of extra costs. The division of risks and other terms agreed in contracts between the partner firms are likely to depend on the firms' relative business strengths at the time. The lessons of difficulties in practice are the same as apply to all commitments between firms: i.e. that it is prudent to foresee potential problems and to agree clear arrangements for these from the start. Each firm involved can then organize accordingly. An initial belief in a joint interest is better tested by an anticipation of problems before commitments to a customer or to others not sharing the risks are entered into.

JVCs may become more common as it is a means of sharing the risks of investments in projects to anticipate new or changing demands. The engineering organization in the joint system and in each firm has to be adapted to the alternative chosen.

Integration of firms

If the need is temporary and mutual, a joint venture or a consortium can be an effective system between firms. It is potentially strong in sharing expertise but risky in its possible divergence of interests. This problem cannot be overcome in a partial combination of firms without amalgamation. A JVC is potentially unstable but is effective in many cases in practice because there are demands that can be met only by joint working which firms accept to be worth some conflict with their other interests.

Experience of joint working in a JVC may lead firms to consider amalgamation. Integration vertically is the means of achieving coherent investment to produce finished products according to predictions of demand. Horizontal integration is the means of achieving flexibility in the use of resources and of spreading risks. Firms can combine related activities included in both patterns of integration. Other firms not included tend to become jobbers, not investing in development but owning the capacity to take on work as the flexible reserve to an industry and likely partners in JVCs.

Further reading

Ellison J. and Kling E., *Joint ventures in Europe*, Butterworths, London, 1991.

Fleming C., in Burbridge R. G. (ed.), *Perspectives in project management*, Peter Peregrinus (Institution of Electrical Engineers), Hertford, 1988, Ch. 5: Management of joint ventures.

Pfeffer J. and Nowak P., Joint venture and interorganizational interdependence. *Administrative Science Quarterly*, 1976, **21**, 399–418.

PART III
Relationships within organizations

7 Case studies

The cases

The preceding chapters indicate that the system of organization within a firm should provide the links needed with each customer, supplier and other outside people, and should enable the specialists in the firm to work together to achieve the projects for the firm, thereby helping their firm to succeed and survive.

Five case studies are described in this chapter. The cases are drawn from a larger number of real instances which have been described in publications or observed in research, but they are presented in a simplified form so that the firms cannot be identified.

The five cases presented are

> a manufacturing firm, denoted by code letter **M**
> another manufacturing firm, code letter **N**
> a firm promoting capital projects, code letter **P**
> a firm of consulting engineers, code letter **Q**
> a contracting firm, code letter **R**

These code letters represent the name of the firms. The diagrams in this chapter are labelled with that code letter followed by a second code letter which denotes what the diagram shows, i.e.

> **F**: a diagram of the flow of engineering information
> **G**: a diagram of the system of grouping of people in departments, sections, etc., in the firm
> **V**: a variant of that system.

Figure **MF** thus shows the flow of information in firm **M**.

In the descriptions of the five cases, the manager responsible

59

for the engineering resources is shown with the title of *Chief Engineer*. Titles vary greatly in practice and are not a reliable indicator of responsibilities and authority. The one title is used here to make consistent reading. In the diagrams of information flow such as Fig. **MF**, the fine dotted lines indicate relationships in the feasibility stage of a project, the broken lines the next design stage, and the continuous lines the completion of detail for manufacturing or construction.

In Fig. **MG** and others in the chapter, the grouping of design and other people in a firm is indicated by means of the type of organization chart evolved by A. K. Rice and colleagues.

Manufacturer M

Firm **M** manufactures a type of machine to the orders of other firms who use them in their factories to make consumer goods.

Nearly every order received by firm **M** is a new project requiring the design of machines to meet a customer's decisions on the performance and previous projects. The firm therefore undertakes many projects, requiring mainly adaptation of design from previous ones.

The flow of information starts in discussions between potential customers and the firm's sales department, at the point marked 1 in Fig. **MF**. The sales department consults the Chief Engineer on customers' requirements and design uncertainties and then indicate proposals and prices to a potential customer.

Design decisions by the firm become more definite at stage 2 when a potential customer requests a tender from the firm offering to supply machines. The customer's enquiry is passed to the Chief Engineer and a draughtsman who works with him on what are called feasibility studies. They prepare the layout drawings, specifications and other details required for tendering, consulting other members of the department if this is thought necessary by the Chief Engineer, but avoiding detail or using the resources of the main drawing office.

The second stage of work of preparing sufficient information for a tender usually proceeds as indicated in Fig. **MF**. There may be recycling in this stage during negotiations with a probable customer. There are also many other relationships between members of the department, such as the managerial ones between the Chief Engineer and all others. Fig. **MF** shows the essential

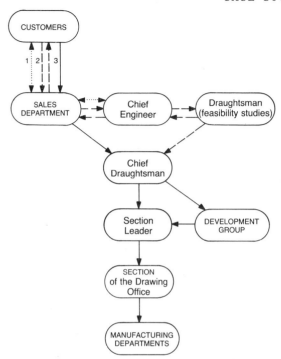

Fig. MF. Flow of design information in manufacturing firm M

links in making the decisions in the design of a project.

Stage 3 is the detailed design work. This is not normally started until the receipt of an order to supply machines, although a few customers do ask to see some layout or system drawings before placing an order. When an order is received by the firm, the authorization to proceed is passed to the Chief Draughtsman for him to plan the remainder of the design of the machines ordered. He decides what can be adapted from previous projects, completes a specification of the performance required of the machines and specifies any development which appears to be needed.

Development is the specialist work of a small group made up of mechanical, electrical and control engineers and technicians. It is separated from the rest of the department as it uses the firm's manufacturing facilities.

Many components for the machines are purchased from other firms, such as motors bought ready-made to standards or bearings

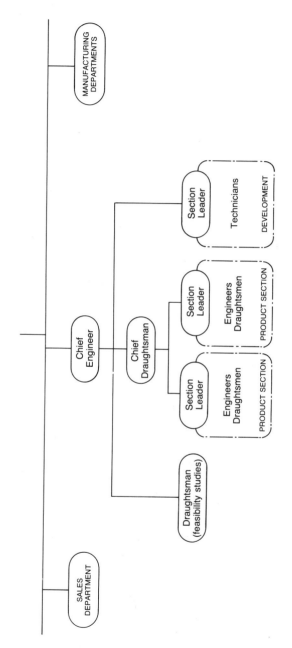

Fig. MG. Grouping in design department in manufacturing firm M

ordered to detail specified by firm **M**. Design problems arising during a project are not shared with the component suppliers. As shown in Fig. **MF**, the decisions in design are completed in this department.

The detail of the design for a project is the work of the main group of engineers and draughtsmen forming what is known as 'the drawing office'. They are divided into several sections because of the numbers of people employed. Most members of the drawing office have been trained in mechanical engineering. The sections are similar and equally experienced in the type of machines made by the firm. A project is allocated to a section according to its total workload. The grouping of people in this example is shown in Fig. **MG**.

This firm is an example of design divided by stage of projects and by level of risk. The Chief Engineer takes part in decisions before the firm is committed. The Chief Draughtsman decides the main sequence of decisions. The section leaders in the drawing office are responsible for the completion of detail for issue for manufacturing. Nearly all the members of department are of the same subject specialism and there is only a secondary subdivision in the small group of development engineers.

This system depends on the Chief Engineer's experience in foreseeing problems. He has the assistance of others in preparing the information for tenders which will commit the firm, but neither they nor he automatically receive a regular feedback of the consequences. Unforeseen problems are likely to be referred back from the section trying to solve them and complete the detail, but the system does not otherwise provide a means of accumulating experience of the consequences of decisions.

The system is flexible in the use of the resources of the drawing office. The resources of the small Development Group appear to be limited, but as the firm is a manufacturer it is likely that more factory resources can be used when needed.

All customers should get the machines they want but have to wait for design and manufacture after giving their order. A customer has to wait longer if he wants the machines proved before delivery. The firm has to make a great variety of machines but usually only a few for each order. This is uneconomic. The firm is also unprepared to provide novel machines if required for new markets or for customers' changing demands.

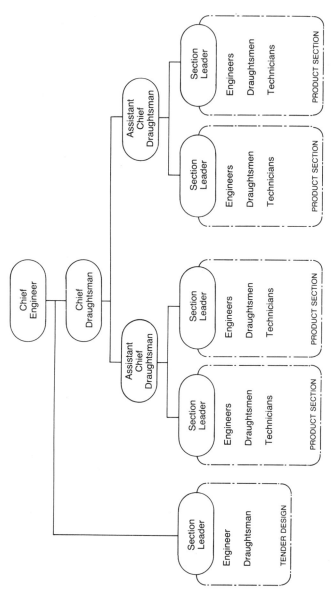

Fig. *MV*. A larger scale of grouping in manufacturing firm *M*

Variant MV

Figures **MF** and **MG** may appear to show a relatively small organization. This is not necessarily so. Its origins may have been small. At the start, the roles of Chief Engineer and Chief Draughtsman may not have been separate and there may have been little or no development work. From this, it could have grown to the system shown in the two figures. It could continue to grow without changing the principles of the system of organization. Greater capacity for tendering could be provided by establishing a section under the Chief Engineer to specialize in this stage of design decisions. Greater capacity to receive orders could be provided by increasing the size and numbers of sections in the drawing office, and an Assistant Chief Draughtsman could be appointed to share the design and managerial work of the Chief Draughtsman.

In a further evolution, the drawing office sections could specialize in types of machine. In this way they could accumulate more experience of some problems, although the result would be loss of flexibility in meeting the load of orders received by the firm. In such a system there could be several Assistant Chief Draughtsmen who partially specialize in design decisions for types of machine, while the Chief Draughtsman concentrates on the management of the drawing office and is concerned with adjusting the allocation of resources between the drawing office sections.

Continuation of these trends might then lead to a division of the development resources according to specializations used on some types of machine. If customers increasingly wanted novel projects, combination of the development and drawing office sections specializing in types of machine could then follow to form product groups, each following the flow of information under an Assistant Chief Draughtsman in place of the one Chief Draughtsman. Thus we could get to the grouping shown in Fig. **MV**, but in principle the relationships would remain as analysed and shown in Fig. **MF**.

Manufacturer N

This firm also designs and makes machines bought by others to equip factories, but it differs from firm **M** in designing and proving prototypes ahead of orders to establish a standard range of machine modules offered for sale. More innovation and development is incorporated. Detail can be varied for each order and some orders to design to order are accepted, but the policy of the firm is that most orders should include only those variations which can be included in the time needed to make and assemble the designed and tested modules.

The firm therefore sets out to sell machines based on modules and sub-systems as already designed, proved and in production. Success in this depends on predicting the demand and proving innovations. Most projects consist of the design, manufacture and testing of a prototype, ending with issuing the final detail for production and assembly for each order.

The sequence of information starts in the firm with internal studies of changes of demand for its customers' products, at 1 in Fig. **NF**. Feasibility studies are coordinated in meetings of a Product Planning Committee, consisting of the Chief Engineer and members of the firm's Market Research Group. The heads of the manufacturing and other departments also attend to comment on proposals and to hear the recommendations made for investment in new projects.

The remainder of design, stage 2, follows the firm's decision to proceed with a project. The decision is passed through the Chief Engineer to the Chief Designer who plans the consequent work. The latter specifies the development required and considers the range of variations from the standard machine proposed that may have to be allowed for to suit customers' orders.

The Chief Designer is responsible for completion of design by a group consisting of a section of design engineers, some specialist engineers and a section of the drawing office specializing in the type of machine being designed. There are several such sections within the Drawing Office which work with the Design Group in that stage. They are later responsible for issuing detail to others for prototype manufacture and testing, for sales tendering and finally for manufacturing. The main design stage is completed when the results of prototype testing are formally reported. This is fed back to the Chief Engineer. Recycling of decisions may follow, either because of unexpected results from a prototype or because of changes in predictions of demand.

After a proved project is offered for sale, stage 3 starts when an order for machines is received by the firm and what is required is passed to the drawing office. The Chief Draughtsman is the link

CASE STUDIES

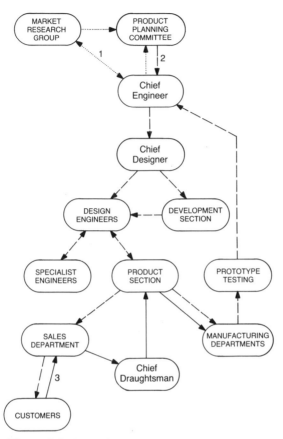

Fig. **NF**. *Flow of design information in manufacturing firm* **N**

in routeing the incoming orders. He is responsible for specifying the variety of variations which can be accepted in orders and is the consultant to the Sales Department in negotiations with probable customers. If unexpected variations are required, these are passed back up to the Chief Designer. In his capacity as manager of the drawing office, the Chief Draughtsman allocates resources to the sections but is not responsible for design decisions until a project is proved for offering as a new standard for sale.

The grouping of people within the design department in firm **N** is shown in Fig. **NG**. Nearly all are specialists in one branch of engineering but within the department there is more

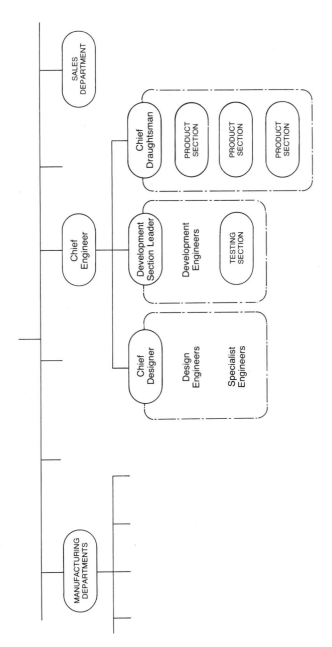

Fig. NG. Grouping of design department in manufacturing firm N

specialization than in firm **M**. Most of the decisions are made in one design group that is only loosely subdivided and is flexibly coupled with drawing office sections concerned with subsequent changes of detail. There is relatively less division of work by stage and there is feedback of results after the main stage of design. Development and prototype testing are separate because they use distinct facilities.

Firm **N** differs from firm **M** in that nearly all design decisions are made in anticipation of achieving sales of machines. Planning has to be the basis of decisions, and innovations have to be proved.

The disadvantage is that this system is not prepared for much design to order, but if demands are anticipated there can be standardization in production and quick delivery of proven machines.

Variants

Differences from the example illustrated in Fig. **NG** would be likely if many projects were in design simultaneously or if the projects varied more in type, in novelty or in regularity of work. As in the previous example, any of these differences might lead to more specialization and division by type of project, especially in the main stage of design.

Product planning could be a greater volume of work and the specialist task of a separate group. If innovation grew in importance, there could also be another group who would study advance ideas based on predictions of longer term changes in customers' markets. Their logical start would be from information on the results of customers' research.

In firm **M**, nearly all design followed a sale. In firm **N**, nearly all design preceded a sale. In the manufacture of capital equipment there are intermediate examples of firms which have to develop and prove a product before sale but have to change much of its design to adapt it to each customer's requirements. In such conditions, the decisions made by the Design Group in firm **N** would have to be revised after each sale. The requirements of each order could be fed back to the Design Group or more design resources added to the Drawing Office for the section concerned to decide the adaptation needed for the order. The latter system is found in practice. It introduces an overlapping specialization in the stages of design, but it is logical if adaptation without further development is required in revisions for each order, or if this stage of work would interfere with the design of new projects.

Firms specializing in types of sub-system may have families of projects on offer that they will design to customers' demands, all of a family being based on a proved process or system but each project for a customer consisting of design of a variant and its auxiliary equipment, services, etc., to meet an order. Manufacturers of computers and manufacturers of engines may, for instance, have a basic set of products offered in this way, but each order is, in effect, a distinct project consisting of design, manufacture and other work particular to customers' requirements. For such work, three categories of design groups could be appropriate: one category comprises groups specializing in basic systems; another category consists of specialists in supporting equipment, services, software, etc.; the third category comprises project engineering groups drawing on the others as required for each order.

For many firms these divisions could be too complicated, but if several basic systems are being developed and revised while also being used as projects occur, these separate categories of groups could be valuable so that the objectives of the commitments to apply a system are not confused with the objectives of further evolution of the systems and of their supporting auxiliaries, etc.

Licensing

To share the costs of the design and development of new projects, firm **N** might sell to another firm a licence to use the details. Alternatively, firm **N** might buy such rights from another firm. These arrangements are common between firms in different countries.

Although the licensed firms would usually obtain all details of the machines as designed and proved, it is likely that the information would have to be reviewed by them and the detail revised to suit national standards or the manufacturing facilities of the firm. An overlap in design between the two firms would therefore be required in these conditions, the receiving firm needing to repeat much of the sequence of decisions in a similar way to the intermediate case just described of partial redesign following an order.

Promoter P

Promoter **P** is a public authority which provides services in a region of the country. A project consists of civil engineering and building construction, and the installation of mechanical and electrical equipment. The operation of the completed installations is almost entirely automatic. Innovations by the suppliers of equipment are utilized, particularly to increase reliability and automation, but the greater decisions in the design of a project are in adapting civil engineering experience to the conditions of a site. Only occasionally is more than one such project in design although there are also many minor projects proceeding to improve or replace equipment in use.

Decisions begin from the promoter's forecasts of changes of demand. The performance and timing for a new project is proposed, together with assessments of when to close old installations which have become uneconomic to continue to use and maintain. Two stages of study and evaluation of new proposals are required: stage 1 to indicate very approximate figures for use in the long-term planning by the national authorities; stage 2 to estimate performance and costs more accurately in order to obtain financial approval for a new project. As in all such work, there can of course be repetition of both stages in trying to fit this promoter's requirements into national policy.

On the agreement to proceed, there follows the main design of a project, stage 3. Civil engineering and building work is detailed sufficiently to invite tenders from contractors who compete for an order for all this work under the traditional procedure described in chapter 5. At the same time, specifications and layout drawings are prepared to invite tenders from manufacturing firms for the supply of sections of the mechanical and electrical equipment, some firms competing to design or adapt design to order, others competing to provide their standards. The result is that a range of firms such as our previous examples **M** and **N** become suppliers to promoter **P**. Detail to fit their equipment is incorporated in completion of the civil engineering and building drawings for construction.

Before construction starts, some details of the proposed structures and services have to be submitted to other authorities responsible for public planning consents and other statutory controls. After construction the drawings have to be revised to show detail as built and instructions have to be completed on the maintenance and operation of the investment.

The flow of information in the planning and construction of such a project is shown in Fig. **PF**. This indicates that the study of increases in the demand for the authority's services and the economy of installations in use is the work of a Systems Planning Engineer. He assists the Chief Engineer in preparing proposals for the long-term planning.

The next stage of design for financial approval chiefly requires decisions on the choice of a site and the layout of structures. This is work for the civil engineering section. Some decisions are required on the performance of equipment. These decisions are made by the Equipment Group.

Other sections take part in the main stage of design of a project, particularly sections specializing in types of equipment, buildings and site work. Sections specializing in operations and maintenance are consulted, in the quite elaborate way indicated in Fig. **PF**.

This parallel involvement of several sections requires planning of the flow of information and the linking of comments with decisions. To achieve this is the responsibility of the section with the most work in the design of a project. Fig. **PF** shows this coordination by the civil engineering section, which usually has the most work because it has to decide details for construction. This section is therefore particularly liable to have a fluctuating workload, rising to a peak in detailing a project for construction but requiring considerably fewer resources when employed only on the prior stage of preparing a proposal for financial approval. The civil engineering section is therefore assisted by a firm of consulting engineers who carry out much of the detail for a project.

Consultants' resources are used when required to take the peak of design work at this stage of projects, but it is also convenient to separate the detail in this way in order to control changes of decisions from the other sections. Geographical separation thus has value. The consultants provide design services and the supervision of contractors on site, but they do not have the independent authority of 'the Engineer' described in chapter 5. The promoter's civil engineering section retains the responsibility for contract administration and the coordination of all decisions.

For lesser projects to improve or replace existing equipment a similar arrangement applies, but with either the mechanical or the

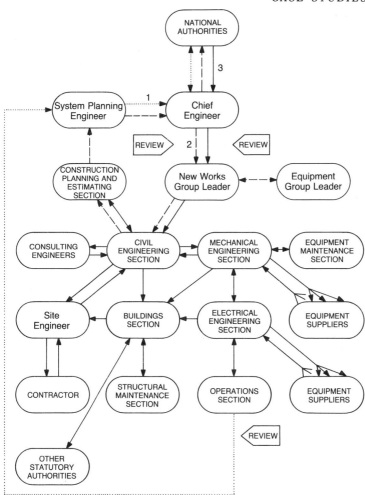

Fig. **PF**. Flow of design information in promoter firm **P**

electrical section acting as the coordinators, depending on which has the primary design decisions.

The importance of planning to meet increasing demand is reflected in the system of organization shown in Fig. **PG**, but with the result that planning decisions are separated from the subsequent stages of design and from the operating responsibility for using the resulting installations. In the sequence of design

PRINCIPLES OF ENGINEERING ORGANIZATION

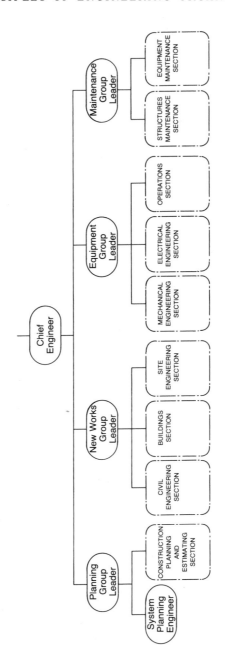

Fig. **PG**. *Grouping of design and related engineering groups in promoter firm* **P** *(N.B. Leaders of section are not shown owing to lack of space)*

decisions there is also a division by levels with division by stage, in a way similar to detail in the previous examples. The main stage of design is divided among sections which work in parallel and with outsiders. The section most involved also acts as coordinator, a choice which is logical if that section makes most of the design decisions and others are dependent on it to fit in their decisions and advice.

Also shown in Fig. **PG** are the sections concerned with operating the completed installations. Operations in this example consist largely of supervising automatic equipment. The maintenance sections are part of the engineering department because they carry out minor projects such as the installation of improved equipment and because much of the maintenance must be planned seasonally in relation to operations. Inclusion of these sections within the department should feed back experience more directly into the planning and design of new projects.

A means of reviewing the effectiveness of the system is provided by regular meetings of the Chief Engineer and the Group Leaders. As indicated in Fig. **PF**, they are not involved in the detail of design of a project. In practice, they act as consultants to the section leaders if unexpected problems arise during detailed design which require revision of prior decisions, but they are principally managers responsible for their groups and the use of resources. These meetings take place quarterly to review all projects which have reached a change in stage of work.

Thus there are three points for reviewing of a project, indicated in Fig. **PF**, the first being held to consider innovations when design begins for financial approval, the second for considering the allocation of resources for the detailed stage, and the third being for review of the results after commissioning of the completed installation. By this means there are formal checks on the system, particularly on the effectiveness of the working of sections in parallel in the main stage of detailed design. But it is a slow method of correction. It can be adequate if the projects require little innovation in design or in the system of making decisions. The system is suitable in these conditions and if the projects do not conflict in the timing of their demands on the resources of the sections of the design department.

Variants

Public services such as provided by promoter **P** vary in the extent that they are controlled by a national authority or are decentralized but subject to some central approvals as in this case. If more design and other decisions were made centrally, the engineering role in promoter **P** could be reduced to the detail to suit each site, operations and the minor and emergency maintenance of installations, these being the essentially local work remaining if the planning, equipment design and operating of projects were centralized.

With growth, the national authority might therefore develop a set of large departments specializing in system economics and planning, equipment design, working with manufacturers to sponsor or to check their innovations, civil engineering and building design, and one with local groups specializing in operations.

If the projects became larger but fewer one might expect a further central department or a set of regional groups to specialize in the detail and coordination of the design and construction of each project. The result would be a difference in size and system from that shown in Fig. **PG**, to divide responsibility between a central authority and local subsidiaries. As in all industries, the attraction of economy of scale results in a tendency for concentration of resources. The potential advantages can be great, but there may be a loss of feedback from the local consequences of decisions imposed from the centre.

Considerable differences from the situation described above are found in commercial conditions when a manufacturing firm is the promoter investing in the construction or replacement of a process or factory. Firms such as cases **M** and **N** are promoters of such capital projects. A new installation is the result, as in a public service, but there may be differences in how decisions are made in the study and design of the project because of different policies in accepting the risks of the uncertainties in forecasting demand, in innovating, and in anticipating decisions by competitors. Decisions by suppliers may also be unpredictable. The firm's decisions are liable to change for all these reasons. Later needs for flexibility need to be anticipated in their initial decisions, and collaborative working between all concerned may be needed from the start of design of a project. It may be necessary to link the firm's departments responsible for planning, marketing, design, development, purchasing, manufacturing, maintenance, personnel and finance in decisions through to commissioning.

Figure **PF** shows an example of the most involved group also coordinating the links with others. This may not be satisfactory if the decisions are more complex and require compromises between specialist interests in order to retain flexibility and accept later changes, or if the project will lead to changes in the future relationships between departments because (for instance) of greater complexity in the manufacturing process and the need for more planning and less local control.

For these reasons, a separate project team may therefore be formed to plan a project and to coordinate decisions, consisting of members from the permanent departments temporarily appointed to act together until that project is completed. Design and project management specialists hired from consultants can be added to the team. If the need is not temporary, the firm can establish a project engineering group to accumulate this expertise. In such a system, cycling of people for temporary membership of a project team would be a way of maintaining links with the other groups.

These are the choices possible in forming a team to coordinate a project, the differences depending on the size and frequency of the projects. More complications arise if the team is required to take some decisions rather than to coordinate decisions made by the established specialist departments. The more vital the project, the higher the level of authority of the people from each department who should be appointed to the team. The more urgent the project, the more that a dedicated team must make the decisions, in consultation with the departments concerned rather than being their coordinators.

A project system may have to be established at two levels, a steering group consisting of the managers who control resources being formed to authorize a project team to proceed with decisions with members of their departments. This transient diversion of authority may be needed to concentrate attention on the speed or the cost of achieving a project. It can also be the means of interrupting departmental relationships in order to adjust them to longer term changes in a firm's operations that are a consequence of a project.

Consulting engineering firm Q

Firm **Q** are consulting engineers specializing in a branch of civil engineering. The firm undertakes feasibility studies, design including detail for construction, preparation of contracts, site supervision and contract administration, when employed to do so by the promoters of projects. The firm has to be prepared to undertake this work when required and cannot plan so far ahead as firms **N** and **P**. The need to act on demand is more akin to firm **M**, but the projects undertaken by firm **Q** vary more widely.

Design for a project divides naturally into four stages, each beginning with a decision by a promoter. Stage 1 consists of discussions of a promoter's ideas and the services which the firm can offer. The second stage is a feasibility study, the extent of this depending on the urgency of the project and the accuracy in predictions of cost required to obtain the finance to proceed. Many projects end here or are delayed awaiting the purchase of the site or the consent of others affected. When a project can go ahead, the next stage tends to be a hurried one of making the main design decisions but concentrating on producing sufficient information with typical details to form a basis for asking contractors for tenders to construct the project. Completion of detail proceeds when the promoter places the contract.

In all stages of work for a project, the firm's planning of design depends on the promoter's decisions. In the main stages of detailing there can therefore be delays followed by urgency. The completion of design may have to be particularly hurried if changes to design are requested by the promoter, but in any case this final stage overlaps with construction in order to adapt detail to information from the site.

Members of the firm are also employed as consultants, arbitrators, expert witnesses to enquiries, lecturers, etc., but the above stages of design are the primary task and are the complex work requiring a system of organization in the firm.

The firm is founded on a partnership of engineers. Partner A is a specialist in design of the type of project undertaken by the firm and he is the managing partner. Partner B specializes in construction problems. The other partners act individually as consultants. All the partners share the task of obtaining new work for the firm.

The flow of information on a project (illustrated in Fig. **QF**) begins in discussions between promoters and any partner, but Partner A takes over to advise a promoter on how to proceed

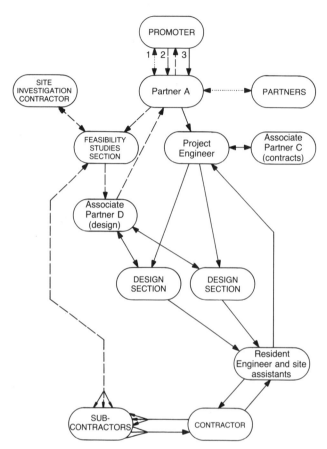

Fig. QF. Flow of design information in consulting engineering firm Q

further and negotiates the agreement to provide the firm's services.

The feasibility studies for all projects are the work of one section of engineers. They specify the information needed from a promoter and from site investigations, analyse alternative proposals and draft a report on their conclusions, working with Partner A in discussions with the promoter but also consulting Associate Partner D who specializes in design problems. They may also need to obtain information from potential suppliers of equipment for a project, from firms who will later be subcontractors to the construction contractor.

When a promoter instructs the firm to proceed with the next stage of design and drafting specifications for construction, the firm assembles a project group of engineers, draughtsmen and assistants to carry out this much greater volume of work. To assemble a group rapidly, temporary use is made of technician engineers and draughtsmen hired through an agency, with recruiting and transfer from other projects following in order that a project group should consist entirely of members of the firm by the time that construction is due to begin. A group is usually divided into two or more sections because of numbers, each under a Section Leader taking an area of the project site as its work.

The group leader is 'the Project Engineer' who is responsible for linking all the remaining decisions until completion of the project. He allocates the work to the sections and is expected to see that decisions are made according to the objectives, programme and budget set in the feasibility studies. This is a testing job. Project Engineers are chosen from all parts of the firm, including site staff. The first ones chosen became the Associate Partners D and E during an earlier expansion of the firm. The present Project Engineers have become experienced specialists in their role, moving on to a new project as a previous one nears completion or joining in the feasibility studies work if there is a delay in projects reaching the main stage of design. For one project, the promoter requested that a Project Engineer should be nominated to work on the project from the start. The firm expects that this may become a common request, but intends to achieve this by attaching the person to Partner A in the early stages rather than by dividing the feasibility studies section into project subsections.

This system of organization uses a Project Engineer to coordinate all decisions on a project but he is not expected to be expert in all the design and contractual problems which may arise. His most important concern is the linking of all decisions in design. The detail is the work of the sections. Their decisions in design are reviewed by Associate Partner D, and their draft specifications and other documents prepared to form the basis for tenders are reviewed by the Project Engineer in consultation with Associate Partner C who specializes in contractual decisions.

The extent of participation by promoters varies in this stage,

CASE STUDIES

some wishing to specify contract terms, to consider the firm's draft documents and drawings, and to issue the final versions to contractors to tender. Other promoters leave the firm to act for them in the traditional way, only wishing to see the tenders after assessment with a recommendation on placing the contract. In either case, Partner A is the link with the promoter until the contractor enters the site.

From that stage, the Project Engineer and his group are responsible for the completion of design and for the administration of the contract, but at the same time continuing their relationships with the specialist Associate Partners C and D.

During construction, Partner A is 'the Engineer', in the UK system described in chapter 5, but he delegates much of this role to his representative on site, namely 'the Resident Engineer', and the Project Engineer. The Resident Engineer becomes the link with the contractor. He passes back to the project group for their comments, drawings and other information from the contractor showing methods proposed for constructing the project.

The system of grouping of people is shown in Fig. **QG**. Partner B has a specialist responsibility for the Associate Partner specializing on contracts and for the site resident engineers and their assistants. The decisions on a project are made by Partner A and the project groups. In addition, an Associate Partner E assists Partner A in planning the use of resources, and in checking the costs to the firm. He also acts as his deputy.

A detail indicated is that a separate section undertakes the main stages of design and contract services for all small projects. This section is grouped with the section specializing in feasibility studies. In practice, they share resources flexibly, according to demand.

The example of firm **Q** therefore illustrates grouping of design by stage, by level and by project, all within one branch of engineering. The specialists in feasibility studies are grouped logically together, as this work depends on expertise in system design, economics and evaluation. Decisions in the subsequent stages depend much more on the conditions of the site and they require recycling of detail as construction proceeds. Grouping by project is then logical when combined with the expertise on design and contractual decisions provided to all project groups by the two specialist Associate Partners.

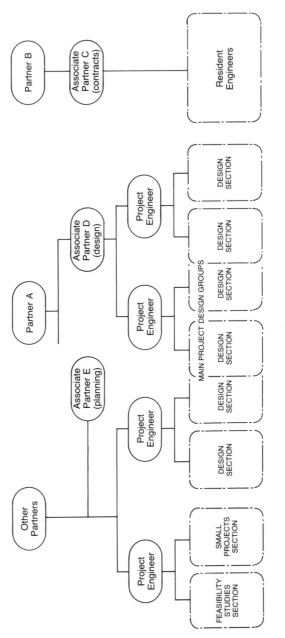

Fig. QC. Grouping in consulting engineering firm Q

The grouping together of all engaged on the detail of small projects is consistent with the system for normal projects and provides the flexibility needed in responding to promoters' requests to study new projects.

Compared with firm **P**, this example illustrates more emphasis on each project as the basis for grouping of people in the main stage of design. This is logical because of the dependence of detailed decisions on conditions particular to a project. It is difficult for firm **Q** to be flexible in assembling project groups when required, but the expedient of hiring temporary employees is used in order to act quickly in response to a promoter's decisions.

The firm's business consists largely of providing these design and contract services. This work is not just a part of a greater interest such as manufacturing. Performance of these services to promoters is the vital interest of firm **Q**. The concentration of decisions by projects is logical in the interest of the firm and natural for the design decisions required in the detailed stages of this work.

Variants

In the above system, firm **Q** is acting on behalf of promoters to coordinate all decisions in the design and construction of projects. This scope of work requires the time of a Project Engineer and Section Leaders, and the continued attention of Associate Partners.

A project could also require firm **Q** to coordinate related work by other firms, such as design by architects and by equipment manufacturers. This would add to the work of the project group, and one possible result could be to increase the functions of the Section Leaders so that they provide these links. One Section Leader could be the link with one firm, another with another firm, etc., although it would usually be preferable on a civil engineering project for the Section Leaders to retain concern for an area of the site rather than to specialize in a type of work.

The contrary could be the choice in work for a building project. For such projects, a firm of architects or consultants specializing in project management would more usually be coordinating all decisions in design, so that firm **Q** would be providing design and some contractual services. Compared with the system shown in

Fig. **QC**, the firm would not need the separate Project Engineers, and Associate Partners might therefore not be employed on such projects.

There would generally be less demand for coordination if another firm were undertaking it, and the same simpler system could be appropriate if the firm were providing design resources only to assist a promoter's engineering department, such as in the example of firm **P**. The result could be fewer people of their level and each leading larger groups of subject specialists.

A similar case might be the system adopted by a design department which was part of a construction firm, and which was undertaking the design in a 'package deal' to design and construct projects. The demand would be more analogous to that in manufacturers **M** or **N**, with a chief draughtsman or coordinating section added to link design with construction.

In a firm of consulting engineers, a more complicated system than the one described in firm **Q** could evolve if the firm continued to expand by undertaking design and construction services for a wider range of types of project. The need and the opportunity to diversify in this way tend to come about because a promoter has used the firm and wishes to employ it to undertake work for other projects. There are potential advantages to a promoter and to the firm of so using their knowledge of the other's work. For this, the firm needs a wider range of subject expertise, particularly for the initial stage of design of projects. To establish the firm in its diversification, additional partners are likely to be required, usually one for each class of project, with consequent changes in the system of organization.

A move into other types of project may be achieved by amalgamation with another firm, so that different systems evolved in meeting differing demands may have to be fitted together and changed later. In this, one choice is for a set of groups to specialize in subjects and one group to specialize in project coordination; or the groups could specialize in a subject but each would also include a section coordinating a share of the firm's projects. With either of these systems, one can combine a continuation of the principle of separate specialists advising all groups and reviewing decisions in design.

Many variants are possible, the choice depending on the range of specialisms required in the firm and on the relative importance of the coordination of each project.

CASE STUDIES

Contractor R

The final example is a firm labelled contractor **R**. This firm undertakes comprehensive contracts to design and supply complete process projects or sections of these, but it subcontracts to specialist firms the detailed design, manufacture and construction of all equipment and structures.

The origin of projects varies. Contractor **R** invests in some research into processes. The results together with market predictions are used to make feasibility studies to present to manufacturing firms who may be attracted by these proposals, and so to promote a project. An invitation to tender and negotiation of an order can follow. But contractor **R** tenders much more often for projects based on processes already chosen by the promoters or requiring the contractor to negotiate to use a process developed by others. In these tenders, the firm is usually competing with other contractors. Once it is working for a promoter, the firm may then get the opportunity to negotiate to undertake further work. As with firm **Q**, contractor **R** has to be prepared to accept orders and to design projects when required.

Most of the promoters in this sector of industry are large-scale manufacturers, and once investment in a project is decided, speed is usually important in order to be able to sell the product and to start to recover the project costs. Development during design is therefore not favoured by promoters, although some may be required to prove the design of automatic control or other systems for processes larger or more complex than built before. Innovations in equipment have to be proved by subcontractors before their products are used. Thus firm **N** might be a typical subcontractor. Contractor **R** needs to keep itself informed of innovations in potentially usable equipment, in order to take advantage of these in offering tenders to promoters.

The start to the contractor's work on a project is therefore usually the preparation of a tender, but it can be in the prior study of a novel proposal to demonstrate its economic value to potential promoters. In either case, the success of a tender in leading to an order is uncertain, and the firm limits expenditure in these early stages of a project to the minimum necessary to estimate the probable cost of the project and to satisfy the promoter that the performance proposed is likely to be achieved. The extent of design completed at this stage varies, depending on the novelty of the process and the policy and knowledge of the promoter. Decisions on the process, control systems and layout of the principal sections of the project are essential in tendering, but the design of structures and the specification of chemical, electrical and mechanical equipment are considered at this stage only so far as they are needed to investigate novel problems or to provide the detail requested by a promoter.

The usual sequence of relationships in design of a project is indicated in Fig. **RF**. The project groups, comprising sections of engineers and commercial and planning experts, specialize in the preparation of tenders. In consultation with all departments, they prepare the programmes for carrying out the design, etc., of a project to follow if an order is received. When available, the person who will be appointed Project Manager may be attached to them during this stage in order to take part in all the planning. As shown in Fig. **RF**, specialists concerned with process performance, safety and layout take part in the design decisions

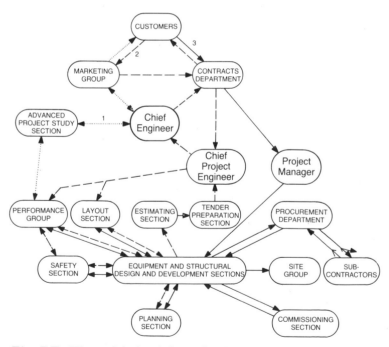

*Fig. **RF**. Flow of design information in contractor **R***

CASE STUDIES

for tendering, but the main group of sections specializing in the design of equipment and structures may only be consulted at this stage of decisions.

This relationship between this main group and the two other groups reverses when an order is received to build the project. The project moves into the main stage of the selection of equipment and the design of linking services and structures. This is divided among the several large design sections which must make parallel decisions. The specialists in performance, etc., are involved only in adjustments of their earlier decisions. Completion of all the work for a project becomes urgent. Changes to the tendering decisions in design of the process should be made only if negotiated with the promoter or if essential to overcome an unforeseen problem. Adjustments to the predicted relationships between parts of the project have to be allowed for as the detail becomes known from equipment suppliers, and the design of structures proceeds to fit the equipment and the conditions of the site. Design of all these proceeds simultaneously, and some recycling of decisions is needed between all the specialists as they complete the selection of the equipment and receive the detail from manufacturers.

In the sequence of decisions, the specialists in sections of a project work increasingly in parallel. They are concerned with detail from other firms which must be linked together by them, and they must meet the project objectives of performance, time and cost already decided in the tender and in negotiations with a promoter. This detailed work is the major demand on the design resources of the firm. Detailed linking and control of decisions is needed in this complex stage of design.

The system of organization needs to combine these requirements with two other provisions important to the firm. The first provision is that the specialists must remain in touch with innovations by other firms so as to be able to consider ideas for new proposals to be offered to potential promoters. The second provision is that at the tendering stage, these specialists must also be available for consultation to the extent needed to study novel requirements and to make sufficient decisions to meet the requirements of the prospective customer and the needs of the firm.

The system of grouping of the people employed in design is shown in Fig. **RG**. There are many sections specializing by subject. These are formed into three groups: two groups consist

Fig. **RG**. *Grouping in design department in contractor* **R**

CASE STUDIES

of the sections concerned primarily with performance and other decisions for tenders; the other and much larger group consists of those involved in the major stage of detail after the firm has an order from a promoter. Thus there is grouping by stage of a project, but sections in the larger group are also expected to make advance proposals for novel projects and they may be consulted during tendering for any project. Work within each of these sections is therefore subdivided in order to make time for the lesser but important demands as well as for the complex detailed work for projects when ordered by promoters.

To achieve this subdivision, each section has a budget which divides its resources of manpower in the way indicated in Fig. **RH**. The proportions allocated to each category of project vary from section to section. These proportions are decided by the Chief Engineer and group leaders at meetings held with members of other departments to look ahead at forthcoming work and to review the system of organization. As indicated in Fig. **RH**, the largest

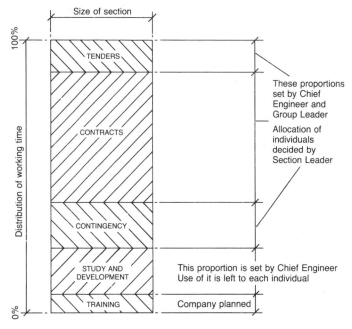

*Fig. **RH**. Resource allocation diagram for a specialist section*

proportion of a section's time is usually planned to be used on design for projects ordered by promoters, i.e. on contracts. Another part of the section's time is for work on tenders. Another part is for the study of innovations and development; the proportion of time allowed for this is set, but the use of this time is decided for himself by each individual member of a section. The remaining part of the section's time is designated as a reserve, to be used as the Section Leader finds necessary for the contingency that unforeseen problems may arise in detailed design requiring additional work.

Figure **RH** shows the allocation of a section's time. The distribution of members of a section to the work for tenders and the work for orders is decided by the Section Leader, but for each project ordered which requires work by a section, one of its members is designated Lead Engineer and becomes the link between the section's work for the project and all others' work for it.

The lead engineers together with a Project Manager and representatives of all departments involved on a project form what is known as the Project Team. The Team is responsible for making all design decisions after receipt of the promoter's order for a project, the lead engineers being empowered to agree decisions on behalf of their parent sections. Depending on the scale and stage of work, the Lead Engineer may be the only member of a section on a project or he may be one of several members on it or he may also be working on several other projects. The Lead Engineers remain at least part-time members of a Project Team until the completion of commissioning checks on the design.

The system is therefore planned to give sections some slack to consider innovations, but, at the same time, to ensure their detailed linking during the main stage of design of a project so that they work within its contractual limits. The flexibility needed to be able to start work on projects when decided by promoters is met partly by allocating people in the sections to work on more than one project so that they can move from one to another as required, the lead engineers providing the continuity of knowledge of each project. Greater flexibility is possible in the amount of detailed design subcontracted to the suppliers of equipment. The latter can be allocated complete sections or subsystems of a project, and the Lead Engineer becomes a coordinator between the subcontractor and work in the rest of the firm.

Variants

One of its subcontractors could resemble contractor **R** in undertaking the design and supply of a sub-system and, in turn, using equipment designed and manufactured by other firms. This subcontractor would require a less wide range of specialisms in design compared with the main contractor. He would also be unlikely to invest in advance projects or in development, but otherwise might have a similar although smaller system of organization.

Greater differences would be likely if this subcontractor were also a manufacturer of one or more of the principal units of equipment required. He would need the resources for designing to tender for the complete sub-system and to manufacture the equipment to be supplied by the firm. For this, the system of organization could therefore well be a combination of elements of the systems needed by an equipment manufacturer (such as firm **N**) and a main contractor (such as firm **R**).

Manufacturers of some types of capital equipment have to be prepared to offer to undertake such comprehensive contract commitments as a means of selling their products. It is a distinct addition to the usual scope and risks of commitment to design and supply equipment. It requires a wider range of design specialisms and expertise in other equipment and structures required to make up the complete tender. Rather than complicate the system of organization required as an equipment manufacturer, a firm may decide to establish a separate department to undertake these more comprehensive design and contracting responsibilities. This might resemble the system shown in Fig. **RG**, but simplified for the scale and variety of design specialisms required. Its relations with the firm's established design work for equipment would be similar to that of the main contractor to any manufacturer of equipment. In effect, the firm's established organization acts as a subcontractor to an added part of its organization, a vertical relationship which may lead to the problems similar to those possible in vertically-linked joint ventures and discussed in chapter 6. A small but separate system for the design of comprehensive projects may therefore be preferable, to keep the decisions on the specification of the equipment required distinct from the subsequent stages of design for manufacture.

Another variant is the type of firm which undertakes the

manufacture of sub-systems using mostly standard components but with some design and development of the sub-system. Such firms could be subcontractors to contractor **R**, promoter **P** or manufacturer **N** to supply them with power, control or data-processing sub-systems. A subcontractor of this type would require specialists in electronics, sub-systems and their applications, but the size and complexity of projects could vary greatly. The system of organization for these specialists could resemble that shown for contractor **R**, the more complex projects being linked by lead engineers in project teams but the smaller projects being the responsibility of the specialist section principally making the design decisions in the way observed in the example of promoter **P**.

Contractor **R** and any of these variants might operate in a joint venture, with others undertaking complementary sections of a project. The effect would probably be to increase the coordinating work required rather than to change the system for the firm's part of the project. If the firm was a member of a horizontal joint venture, then the increasing demand would most affect the Project Manager, and in the same way as might occur when the consulting engineering firm **Q** had to work with others. The lead engineers in contractor **R** could also act as links with horizontally related work by other firms.

In the more complex conditions of a vertical joint venture the linking would demand correspondingly greater work by those normally involved. A vertical link with the promoter would require more detailed coordination through the Chief Engineer and Chief Project Engineer in the early stage of decisions. The alternative of a vertical link with a subcontractor would require more detailed coordination through the Procurement Department. In both cases, the vertical link could be through an enlarged project team with sufficient members to coordinate the detail. A wider range of decisions has to be made as part of working in a joint venture. The expertise and responsibility for these must be added somewhere, as discussed in chapter 6. The choice depends on the uniqueness and importance of the project relative to the remainder of the firm's interests.

Comment

The five cases illustrate the choices in specialization by the stage of projects, by managerial level, by subject, and by project. One case also illustrated an advantage of geographical separation. The cases indicate how specialization can be combined in various ways. Any of the types of firm mentioned might have working relationships with each other for a project, but their systems of organization would be likely to differ because each has its own sequence of decisions to make for its part in the project.

Some of the detail shown only in one case could also be applied to the others. For instance, the experienced engineer set apart to review design decisions could be of value in more instances than the consulting engineering firm **Q**. The meetings to review the system of organization mentioned in other cases could be appropriate to all firms, especially if held when each major project is entering a new stage of work. This is also the case with the scheme for dividing specialists' time between present commitments and future proposals, as described in contractor **R**. Planning and cost control of design can also be separate specialisms. Any of these features could be of value in a system of organization because of growth in the scale and complexity of design.

The cases include instances of the separation of a group given the slack to consider ideas for future projects ahead of demand. Members of such groups can be temporary, to feed forward their knowledge of a project and to feed back their experience of the consequences of decisions. Such a group is likely to be of value in firms in which most design follows after an order from a customer, as the urgent and short term decisions then required may leave all concerned with no time to keep to good intentions to set aside some time to look ahead. Such pressures were likely in firm **M**, where the introduction of new ideas and the results of research depended on the Chief Engineer. Much the same applies to the Associate Partner D in firm **Q**. Their roles need time free of short-term problems.

The systems illustrated vary greatly in the linking between the design of projects and the subsequent stages of manufacture or construction. In a firm which is designer and maker, the cases illustrate instances of a drawing office providing the link. In all firms there could be a need for a chief draughtsman to plan the recruitment, training and allocation of draughtsmen, and to manage the supporting services such as printing, but in the manufacturing firms the person in this role also coordinated the completion of detail to link with the subsequent stage. A civil engineering contractor could require the same to plan and carry out a comprehensive 'package-deal' contract. In the other examples, the detail diverged from a section to a contractor or subcontractor.

The systems described had one person at the head of design and others responsible to him in a hierarchy of authority. In most cases, the title 'Chief Engineer' was used to label this role. In the consulting engineering firm **Q**, the Managing Partner was the equivalent. In practice, titles vary from firm to firm. Common to all these cases was the role of a head of design linking a decision to sanction a project with the system for proceeding with its design.

In every case there was also a change in the level of person in the hierarchy making the decisions at a change of stage of design of a project. The Chief Engineer took part in decisions to commit the firm to design a project, but once the decision had been made to risk investing in a project, the continuation of design was passed down to others. The same tends to happen within any group in a department: the leader is the link with incoming work but he does not have the time or perhaps the expertise to analyse all the consequent problems or to be the link with others working on the related problems.

Leaders such as the Chief Engineers in the cases are least of all likely to have the time to evolve new ideas. They are members of the managerial system and their role includes the planning, recruiting, training and supervision of their group and of subsidiary sections. In the cases described, it might appear that the members of some groups were making decisions without their seniors being involved in the prior decisions. The Equipment Group Leader in firm **P** appeared to be an instance of this, but it was noted that he took part in the reviews of projects entering the system and his role was thus one of planning and of being consulted on unexpected problems during the consequent work.

In the manufacturing industries, many firms are similar to case **N** in that they design products in anticipation of demand. Much of the design and development have to be completed before an order is received. Resources have to be invested at the risk of the firm on the basis of predictions of the demand. Most of the experienced and specialist members of the design departments in such firms work on design ahead of demand, on supporting development and on decisions of detail up to the completion of prototype testing. Comparison with case **M** shows the difference if nearly all design is adaptive work following an order.

The proportion of the design of a project which precedes an order is thus an important variable influencing the systems of organization in design departments. At one extreme there are the jobbing firms working entirely to order without developments to attract demand. They must have the capacity ready to apply on demand. Their risk is that others with developed innovations may get the work. At the other extreme there are the mass-production firms offering products designed and developed in anticipation of demands. Their risks are in anticipating demand correctly and in having the quantity of products available where and when required. Between these extremes there can be any intermediate condition of some design and innovation ahead of demand and completion of the remainder following a sale. An order received from a customer changes the risks to the firm and there is a change in the consequential level of decisions in design.

In every case, the later stages of design included instances of horizontal linking between a section working on a project and the other specialists involved. In some instances this linking was the responsibility of the section most involved in the design decisions, but others illustrated that this linking can be the role of coordinators added to the system. As demands become more complex there is greater need to define who is responsible for the coordination, economy and safety of each project.

Continued evolution

The case studies were observed over a short period. Continued observations might show that manufacturing firms akin to case **M** are liable to have to change to designing products in anticipation of demand, in order to offer faster delivery and to economize by standardizing production. If so, their systems of organization might change towards that seen in case **N**.

A different change might be found. Firms such as case **M** can alternatively get work by becoming manufacturing jobbers, making part or all of machines designed by others. Jobbers need only the design capacity to adapt detail for manufacturing, and so might have a relatively simple organization. Many such firms exist, specializing in manufacturing processes rather than in types of product, and obtaining much of their work as subcontractors to others such as cases **M**, **N** and **R**.

From its prediction of future needs, a firm such as manufacturer **N** or contractor **R** could wish to take over a specialist subcontractor or supplier such as **M**. The consequent system for linking their engineering organizations needs to be designed to combine the distinct demands of the prior design decisions and of the detail utilizing the specialist expertise. Otherwise the potential advantages of the union may not be realized.

Further reading

Rutter P. A. and Martin A. S. (eds), *Management of design offices*, Thomas Telford, London, 1990.

John Hall Boilers Ltd — Case study, Cranfield School of Management, 1971.

8 Specialization within firms

Departments
In every case described in chapter 7, the firm was divided into departments, and the larger departments were divided into subsidiary groups, sections, etc. This is common in government offices, organized religions and military services, as well as in industry. If a firm is small enough for everyone in it to work together in one room they can share tasks and information informally and change how they do so day by day and hour by hour. Most firms are larger than this; many are much larger. They typically have systems of organization in which departments have different functions, and within these departments, groups and individuals specialize in more detail.

Various ways are possible and logical for how they can specialize. One common way is the stages of the work for projects. All five cases illustrate this principle to some extent. In such systems, one group of people can specialize in feasibility studies and hand over to another group the consequent design of selected projects, as in case **M**. Following this principle, a third group could be the specialists in detailing for manufacturing and construction, as in case **N**. The middle group would be the specialists in analysing design problems, but they would have no links with the demand or with the subsequent work to complete the projects.

Levels of experience form another basis. All the cases illustrate the way in which decisions are delegated down a managerial hierarchy as the work proceeds. Another basis is the expertise needed in analysing problems so that people can be grouped according to their specialist knowledge which can be drawn on when considered necessary for the making of decisions. This was the dominant principle in cases **P** and **R**.

Each way of dividing the work in a firm has some merits, but a consequence of any system of specialization is that it is used as the basis for locating people. The tendency to group specialists together is natural when they share the use of facilities such as experimental workshops or drawing office services, but their resulting separation can greatly reduce communications with other groups. Any division of work between groups of people can hinder the cooperation needed for a project. This combined with geographical separation can prevent the exchanges of information needed to foresee and solve problems.

But why not group people by projects, putting together all the people whose work contributes to the primary task? By following this principle, each project should have coherent attention. Flexibility and regular reform of the organization would also be possible because each group has only a temporary job. This therefore appears to be an attractive principle. Yet it is rare in practice, except as a secondary feature in complex systems which are based primarily on the other principles already described. To know why this is so, the merits and disadvantages of each possibility must be analysed. Reasons for different choices can then be considered.

Grouping by project

The principle of grouping by project is clearly logical as a means of concentrating on the firm's objectives, as projects are the productive work and all other activities have only internal value. It puts together the people dependent on decisions in design, and it should therefore achieve a coherent response to demands and to new problems arising during the work. Applied to all of a firm, it puts together the responsibility for a project and the resources needed to carry out these responsibilities.

It is also the principle that tests group leaders the most, as each for his project has the range of responsibilities of a general manager. This can be a useful part of a firm's scheme for management training.

On the other hand, the needs of a project for various skills, etc., are only transient and they may fluctuate. Sharing of these needs between projects rather than having self-sufficient project groups may therefore be more economical. Experience can then also be shared and reserves of resources drawn on when required.

Grouping entirely by project is therefore a principle that is appropriate for a 'one-time undertaking ... that is infrequent or unfamiliar ... complex ... and critical', as concluded by C. J. Middleton in a study of some examples in firms in the USA. Except in these conditions, it is more logical to form temporary project teams from members of established specialist groups, as illustrated in case **Q**.

Grouping by subject

On the basis of the principle of grouping by subject, people can be grouped according to the main branches of technology such as civil engineering, metallurgy, etc., or in groups specializing in subsidiary branches of one of these, such as soil mechanics, structures, highway design and others within civil engineering. As noted previously, this was dominant in cases **P** and **R**.

An illustration of the detailed choices possible in following this principle for the division of work is given in Fig. 8A. This is derived from the observations of a Netherlands management consultant. It is a polar diagram of the choices possible within a group working on the design of bridges, drainage and road details for highway systems.

The vertical variable is the extent to which people in this group could specialize on bridges, or drainage or road detail. The horizontal variable is the extent to which they specialize in analysis, drawing or in the preparation of specifications. With the greatest divisions of work, a member of the group could be specializing in the analysis of bridge problems. With the least divisions of work, all members of the group could accept any part of the work. A mixture might be best, particularly to be flexible, with coordination of the use of the group's resources and the results for each project.

A diagram such as Fig. 8A can be used to compare the predicted demands of projects with the expertise and interests of individuals in a group, and so to define needs for recruiting and training.

Specialisms can also evolve which draw together parts of several branches of expertise for application: control engineering is an example.

Specializing in a subject and becoming expert in its application is the way in which 'know-how' can be accumulated, the lessons learnt from mistakes and the experience of similar problems shared. All such specialisms are learnt by study and application. Further study and meetings with corresponding specialists in other firms to exchange experience in that branch of technology are necessary for individuals to keep up to date in their subject. The firm advances its expertise by investment in research, by individuals keeping up to date, and by recruiting.

Members of subject groups can act as reserves for each other and train new members so as to form a pool of expert resources. Grouping by subject is therefore the logical principle for establishing expertise for application among projects when required and for accumulating experience from one project to the next. It is suitable for linking design with development work in the firm or by outside specialists.

Specialism by subject is vital in engineering to enable individuals to master and to advance sections of the expertise needed in the design of projects, and it is therefore a feature of most organizations. This principle offers careers in specialization. Fewer demands are made on group leaders if their task consists of the types of problem

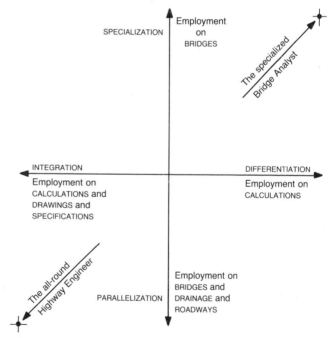

Fig. 8A. Choices in specialization for each person in a highway design section (after J. S. Hulsoff Pol)

familiar to them which arise in their group's specialist work. If they are also required to resolve conflicts between project and specialist objectives, the demands on them are greatly increased.

The disadvantage of this principle is that work divided among different specialist groups has to be fitted together to meet project objectives. There can be confusion between them because objectives vary from project to project, and there can be conflicts of interests because the responsibility for results is indirect. In the solution of a design problem, a specialist may wish to spend more resources to reduce the uncertainties in his predictions of the consequences of choices, whereas other people concerned may consider that a decision must be made and some risks accepted because of limitations on cost and time. At its extreme, this is a conflict between the safety of the fully analysed solution and the reality of economy and consequent risk. Such conflicts tend to arise when a project is divided among large numbers of specialist groups. Any one group can regulate only its own part of the work. Their specialism may be their dominant interest. To them, project objectives may seem uncertain and can become secondary in influencing their decisions.

Division of the work for a project among specialist groups and individuals can lead to failures of communication because each specialism in engineering and in other professions and crafts tends to develop a different language and have only a limited knowledge of others' work. These problems can occur on small as well as large projects. The resulting loss of cooperation and of flexibility in the use of resources can be uneconomic and demotivating. Hazards may also be overlooked. Interest in their specialism rather than its contribution to the ultimate success can become the objective of a group or an individual specialist, as observed by L. B. Barnes in US engineering departments and since reported from many cases. Conflicts of interest can develop between individuals, firms' long-term objectives, the immediate opportunities, ethics and law. Failures in understanding and communications can be greater when the people in a firm work in departments, sections, etc., in different parts of a building, and more so if further apart.

The least division of work among groups of specialists is logical to minimize these risks. It seems obvious that one should try to locate together all who are dependent on each other's work or on common information. The potential disadvantages of separating people into subject groups may be acceptable if the division of knowledge of a project as a whole is required to achieve secrecy or security. What has been called coupling (Stinchcombe and Heimer) between groups is required to achieve the more usual industrial need for collaboration in solving problems.

Grouping by project stages

In the alternative of dividing work in series, groups of people can specialize in a stage of decisions with an immediate objective in the evolution of each project. Chapters 4 and 5 describe vertical divisions in the relationships between firms. Within firms the same is also to be found, as illustrated in the case studies. In manufacturing and construction firms, different departments specialize in design and in production. In contracting firms, different departments specialize in preparing tenders and in undertaking the consequences if a tender leads to receipt of a contract to carry out the project offered. In such systems, experience is accumulated in each stage. People with similar prior training who enter different departments become experts in different stages of work.

Within the design stages of projects such a division of work can be flexible. One common arrangement, mentioned at the start of this chapter, is to follow this principle by allocating to one group the initial stage of analysing demand, research results and ideas up to the point of deciding whether or not to proceed with the proposals. This group thus becomes the specialists in feasibility studies. Its output is the input to the next group.

It is noted in chapter 3 that slack is required for innovation. The initial stage of studies of ideas can be separated for this purpose, freeing one group from the detail and the urgency usually required in the subsequent design of the projects selected. In this way, they can have the time to define new objectives and to look for new ways of utilizing the firm's resources. This stage can be subdivided, so that one group is free to consider innovations and another group to specialize in the evaluation of proposals. This system can be appropriate when the firm is engaged on innovative and on adaptive projects, to separate the initiation of these two types of proposal. If the demand for one of these classes of work fluctuates, then only a few people need specialize in it and others from outside or from other groups can be allocated to assist them when necessary.

Division of the work by stage can be applied to the remainder of the design sequence. After the selection of a project, its design can become the responsibility of a group specializing in the main stage of decisions. This group would logically be the link with development testing. As in case **N**, a further group could inherit the results of their decisions and be the specialists in detailing and preparing the other instructions for manufacture and construction, and the subsequent stages including information on how to use and maintain the product.

On the basis of this principle, the first group has the objective of assessing the choice of projects, the second the solution of design problems and the third the issuing of the results. From assessment of *whether* a project should be chosen, the work changes to deciding *how* best it should be realized and then to specifying *what* needs to be done to make it. By division of the work into these three stages, each group has an immediate objective. The system reflects how the nature of problems and decisions changes as the work proceeds.

The disadvantage of this principle is that it depends on communications from stage to stage. The acceptance of prior decisions may not be successful because of failures in one stage to foresee problems that arise in a later stage or because of human uncertainty about accepting the risks of inheriting the consequences of decisions made by other people. A system of organization based on the principle of division of work into stages is therefore most likely to be effective where projects require only adaptive design and when the system has become familiar to many of its members.

In such a system there should be regular cycling of people forward to earlier stages of the work so that experience of consequent detail is renewed in the groups making the main decisions in design. In some cases this is possible because people can also move with a project through the stages to work with others specializing in a stage. This is not possible in many other cases where the continuation and the timing of projects depend on uncertain changes in demand.

Stage by stage grouping forms a natural basis for the control of costs and progress, but any such system depends on the capacities of the groups being planned to meet the flow of work from one to the next. Most of the problems of the group leaders are repetitive ones arising in the linking of the stages.

Grouping by levels

Figure 3A illustrates the characteristic pattern of the sequence of problems in the design of a project branching downwards from an initial critical decision. The higher the level, the more abstract the thinking required. As decisions proceed they become less profound and more predictable. Systems of authority for making decisions can be based on this pattern so that the initial choices are made by people at a high level who then hand down to others the authority to solve the consequential problems. Delegating authority at a series of levels to form tiers in a hierarchy is an ancient expedient of organization.

The sequence of decisions to be made in the design of projects would therefore be a logical basis for defining authority to make these decisions and it would probably be widely understood in practice. This is partly demonstrated in examples of systems of organization observed in design departments in firms. Division of decisions by levels is found at the initial stage of design in firms and within specialist groups, where the group leader takes part in analysing and solving the initial problems in the group's work for a project but delegates the consequent detail down one or more levels.

As cases **M** and **N** show, divisions by level are most often found combined with divisions of the work into stages.

The disadvantages of division of work by levels are that it is inflexible in utilizing the abilities of individuals and that it does not establish the links which are needed between related choices for solving parallel problems in the sequence of decisions. As with the associated principle of division by stage, division by level is likely to be effective only in settled conditions of adaptive design. It may therefore be of diminishing value in practice as decisions in design get more complex and uncertain.

Functional grouping

The term *functional* organization is used in textbooks to mean in effect grouping by subject and by stage combined. The word is avoided here so as to keep clear its two principles, and also because it is used to denote a principle of supervision which is referred to in chapter 9.

Drawing offices

The greater uncertainty of project decisions has changed the role of the 'drawing office' in many engineering firms, particularly in the manufacturing industries. In British firms, the drawing office was the design department, consisting of engineers and draughtsmen under the Chief Draughtsman, and responsible for issuing all the design information as instructions to the factory. This was so in case **M**. Nearly all the members of these drawing offices had been craft apprentices. Their work was largely adaptive design and their craft training was used to make decisions to suit manufacturing.

This has changed, as more analytically trained technologists have been required to assist drawing offices in the analysis of possible solutions to novel problems. In some industries they tended to be employed initially as specialist advisers to the drawing office and to be placed in a distinct area or 'technical office'. This separation of specialists may have been natural because they required experimental facilities or because they were not experienced in detailed design. It has become less natural, because of the tendency for the problems in design to become less dependent on adaptation and because the initiation of projects has become more dependent on abstract analysis. What had been the advisory 'technical office' has become increasingly responsible for making the design decisions, as in case **N** compared with case **M**. Where separation has been retained, the relative positions of the two offices have changed, creating a division of authority between the main 'design office' and the drawing office.

Such divisions may have evolved naturally. They are not necessarily logical. Problems of motivation, morale and innovation can result. They may be best answered by mixing rather than by retaining a division by levels, so as to give opportunities for individuals and to make flexible use of talents in the solution of novel problems. The increasing novelty of problems appears to have been the cause of such divisions in design, yet the evidence is that divisions of work should be reduced in such conditions.

Separation for cost control

One argument for dividing a department into specialist groups is that each group is then a distinct service and its costs can be controlled by projects having to request its services formally. This assumes that the specialists would work effectively in this way. It also assumes that whoever needed their services would be willing to call on them in good time and not be reluctant to do so because the system required justification and accounting for time used. Such formal separation would be logical if the specialists were very rare and expensive. This is not generally so.

Accounting for much of the use of people's time is achieved in most firms by their recording the hours spent on each project or part of a project. Control so that resources are well used is not achieved by such means. Accounting tells the tale too late for remedying the results. Planning is the means of using resources well. Separation of specialists is logical when the problems are predictable, but less and less are design predictions reliable. Design is not manufactured. It is probabilistic. Only on this basis is some prediction possible. The evidence for and against controls is very limited, but it is clear that the trend in all industries must be for employment of a growing variety of specialists mixed so as to pool their ideas, experience and knowledge.

Further reading
Martin A. S. and Grover F. (eds), *Managing people*, Thomas Telford, London, 1988.

9 Coordination and control

Managerial hierarchies
Locating together everyone in a firm who is contributing to a project so that they know each other's thoughts and decisions may not be practical for reasons of numbers. If they are separated, they need to be coordinated. Choices and traditions in dividing work among specialists in firms are commonly learnt from experience. Choices in achieving coordination are equally important. They are not so readily learned from experience, because they are more apparent when failing than when succeeding.

The problems of coordination in organizations are not new, especially those due to size. As far back as in the pre-Christian Old Testament referring to events 3000 years ago descriptions can be found of a society divided into groups with intermediate leaders because its problems were too large for resolution by one person alone. A hierarchy of authority under a 'top' leader is clearly described, with several levels of intermediate leaders. This principle is nowadays commonly accepted in industry, and it is also common in social organizations. It is useful provided that there is agreement on the rules for the use of authority and the capacity to manage the unexpected.

Every case described in chapter 7 illustrates the traditional way of achieving this by leaders of the departments, groups, sections, etc., forming a hierarchy of managers under a chief executive, government minister, general manager or managing partner at the 'top'. Fig. 9A illustrates the principle. The top manager receives resources and the authority to use them from company shareholders, partners or central or local government, and is responsible to them for the results. Organization charts or 'organigrams' of this form are commonly used in firms showing the names of

PRINCIPLES OF ENGINEERING ORGANIZATION

departments and the managers' job titles. Positions in the hierarchy indicate who is formally responsible for each division of work and for its coordination with the others.

In Fig. 9A, abbreviated titles are shown, the top job being labelled 'CE' to indicate the title of chief executive, the next level 'DM' to indicate senior managers with titles such as divisional manager, department manager or director and manager, and below them middle managers, first line supervisors, and the employees who produce the firm's products or services.

As the cases show, in practice the number of levels of management and the number of people responsible to the manager above them vary greatly. Fig. 9A is only an illustration of the form of a hierarchy. The numbers of roles and levels in it have no significance.

A managerial hierarchy is a system of authority, essentially over expenditure and other decisions on resources. Much of our language

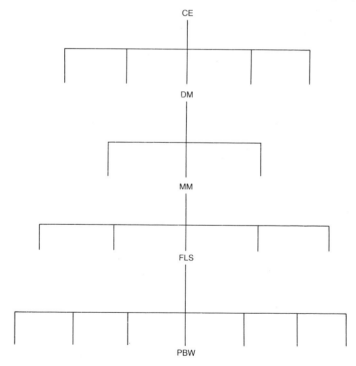

Fig. 9A. Form of a managerial hierarchy

for discussing any system of organization is based on the concept that authority stems from the top of a pyramid-shaped hierarchy. It is an influential concept, although it varies in its effect from culture to culture and generation to generation. It corresponds in form to the branching pattern of decisions in design depicted in Fig. 3A.

There is usually most formality in defining authority at the higher levels, because of longer term risks of decisions, but the least formality between the people in these roles. The lower the level, the more that authority is indicated by titles and the more specific are the decisions.

Delegation of authority

No one person in the role of Chief Engineer in the cases described in chapter 7 could have the time to take part in most of the decisions of the people in his department, nor would he have expertise in all their work. The theory of hierarchies is that the authority to make decisions is delegated down and divided level by level, but each manager remains responsible for the results of subordinates' actions.

By applying this principle at all levels, the higher managers should be free of detail and have time to plan ahead. This is the theory. It is assumed that the consequences of decisions will be fed back by regular reporting, the reviews of projects as in case **P** being one means of arranging this.

In E. F. L. Brech's phrase, the system is a framework for management and leadership. Consultation, counselling, meetings, committees and systems for appealing against decisions are added, depending upon social norms, the number of people involved and the nature and the variety of the decisions.

External commitments

The level of authority in a firm authorized to enter into a contract usually depends on the size and risks of each contract. The responsibility for the management of the consequent commitments is then usually delegated down the hierarchy, as illustrated in the cases. This is the practice in customer and supplier firms in all industries. It is expected to be much riskier to enter into a contract than to carry out the obligations in the contract. Delegation of these follows the principle that lower levels of management are authorized to take less risk.

On agreeing a contract, the authority to communicate with the other party is delegated typically two levels down. This is logical in the buying and in the selling parties' organizations if they are in their usual situation of having several contracts for one or more projects. A manager authorized to enter into a contract delegates the continuing commitments to a manager of a group responsible for all the agreed contracts, and, in turn, that manager delegates a contract to a member of the group.

Complexity and risks

The common requirement for a hierarchy of management to be effective is that most problems must be foreseen so that the authority to solve them can be delegated down in an understood system. Continuing growth in the complexity and the economic risks of projects will increase the proportion of unforeseen and uncertain problems. It becomes harder and harder to define a project at the start and the subsequent decisions delegated down as less risky.

Decisions are therefore increasingly likely to be recycled during the design of projects, as illustrated in Fig. 2F, requiring the 'organic' relationships between specialists discussed in chapter 8. As observed by D. D. Dill and A. W. Pearson, coordination also has to be organic and demands leadership.

One remedy is to reduce the number or the variety of decisions to be made by a manager. This can be achieved by increasing the number of managers and so decreasing what is called their 'span of control'; but this has the obvious result of increasing the number of managers, the volume of communications required between them, and therefore costs.

Observations by J. Woodward in a range of manufacturing firms showed that the spans of control and the number of levels in managerial hierarchies do tend to vary between the various branches of industry and to relate to differing needs to plan ahead and to make uncertain decisions. As the trend is for these needs to increase, systems of organization based only on the concept of a hierarchy can be expected to have to increase in their numbers of levels and total numbers of managers; but not to much advantage. Only a small reduction in the average span of control in an organization requires a disproportionate increase in the number of managers. The cost and the vertical complications increase and at best there is only a small gain in the effectiveness of the system.

Deputies

In many public authorities in Britain it has been practice for the heads of departments to have a full-time deputy who acts for the head in his absence and is formally the channel of communications between the head and the members of the department. Although informally most heads communicate directly with other members of their departments, by this means they should be free of detail and have the time to concentrate on policy and to think about the future needs, objectives and risks of their projects. A potential advantage to a deputy was experience which could lead to future appointment as a head of department.

In a case in a municipal organization reported by A. S. Martin, the Director of Technical Services* had two deputies. This was done in order to keep the spans of control of the Director and each deputy to fewer than six. The organization chart is reproduced in Fig. 9B. The expected advantage of this was that the Director should have time to concentrate on relationships with Council committee chairmen, other heads of departments and central government officers, while each deputy would manage three or four sections in his part of the department.

In practice, the principal engineers shown in Figure 9B at the third tier level who carried the responsibility for the work of their specialist sections felt remote from the Director. To overcome this, he involved them, as well as one or both deputies, in discussions and decisions. As a result, it became difficult for the deputies to retain a distinct role with defined authority.

Eventually, the deputy posts were phased out. The third tier became the second tier, responsible to the Director and designated as Assistant Directors. Councillors and the staff in other departments were encouraged to contact them directly, and they carried out their own correspondence — instead of the traditional procedure of all communications through the Director (or deputy in his absence). Together, these changes in responsibilities were very motivating to the principal engineers. The span of control of the Director of Technical Services increased from three to eight. His consequent workload was manageable by practising extensive delegation of decisions, and he was thus able to concentrate on those things he alone could do.

---

*The Director of Technical Services in this example is equivalent to the Chief Engineers in the case studies described in chapter 7.

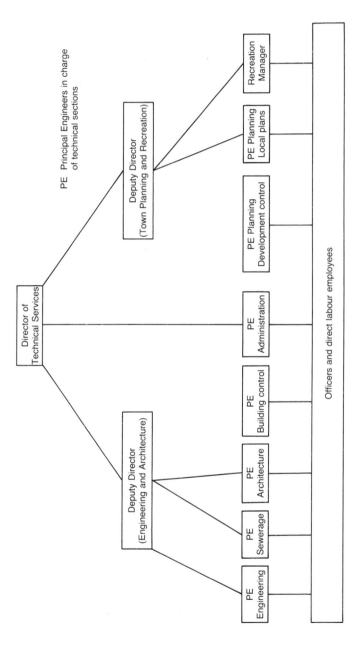

Fig. 9B. Head of department with deputies (A. S. Martin)

A deputy is needed in the absence of any manager, but this case indicates that it may not be a separate job.

Functional supervision

An alternative evolved by F. W. Taylor for the supervision of manufacturing work is to delegate a line manager's authority among a set of supervisors who are specialists in each 'function' being supervised. This idea of *functional supervision* as it might be applied in an engineering department is illustrated in Fig. 9C.

There is no sub-grouping under the functional supervisors. All of them give instructions in their specialist 'function' to all the designers.

The potential advantage is that the total number of managerial roles should be less than if there were sub-grouping under the supervisors. The obvious objection is that the decisions and advice given to a designer by two or more of the supervisors may be inconsistent and have to be coordinated by the chief.

It has been said that this sort of system never became popular because of the '... deep-rooted conviction of most industrial managers ...' that no one can work satisfactorily under more than one boss (J. Woodward). Perhaps not, but convictions tend to be most deeply rooted when not based on evidence. The practice common in industry that managers must heed safety advisers is close to that arrangement.

Separate coordinators of projects

Figure 9A indicates that each person has only one boss. In theory, the sequence of decisions for a project could therefore be the basis for defining which manager should make each decision. The case studies show that systems in practice are more complicated than this, with multiple relationships between various levels and only indirect or occasional communications between some managers and their subordinates.

All of the instances described in chapter 7 illustrate the augmentation of hierarchies to combine specialization with temporary coordinating responsibilities for projects. Each section or individual in such systems can in effect have more than one boss, although having only one in the established hierarchy. In theory, the line boss takes part in the prior decisions to commit resources to a project, and the coordinators make the remaining

PRINCIPLES OF ENGINEERING ORGANIZATION

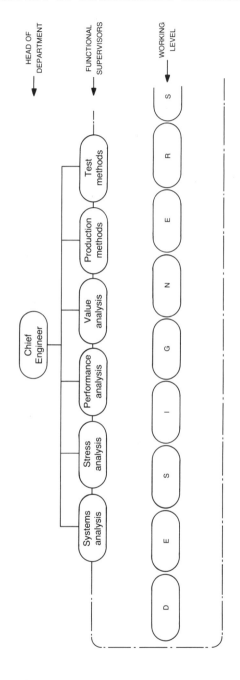

Fig. 9C. Principle of functional supervision (after F. W. Taylor)

decisions, or make sure that they are made. Formally, the coordinators can achieve this only so far as proves possible without having to go back and recycle the prior decisions higher in the system. The level of the coordinating role should therefore be no more than one level below the line manager responsible for the economy and safety of the consequences.

Several of the cases show that the decisions to start a stage of design proceed down a hierarchy, but the decisions during the subsequent stages are coordinated by someone who is not the line manager of everyone taking part. In case **M**, the Chief Designer coordinated design with development, although he was not the manager over both groups involved. In case **N**, the design section had a similar role. In case **P**, the civil engineering section was the coordinator of several other sections. In case **Q**, project engineers were responsible for the design of their projects but made decisions jointly with specialist Associate Partners who were at a higher level in the hierarchy but not with formal authority over the decisions.

In case **R**, several features as in the above were compounded for the more complex linking for each project needed between groups in the firm and with outsiders. In that case, the Project Manager coordinated a team drawn from all the specialisms, making all but the critical decisions without involving their parent section leaders or higher managers. The role can include being the link with the customer or with suppliers. One person might have this role for several projects, or a team might be needed to coordinate a large project.

Line-and-staff roles

The 'line-and-staff' principle is one means of adding a project coordinator in what is called a *staff* position. In this, the *line* is the hierarchy of managers who have formal authority over the resources for the primary task. An organization tree illustrating this is shown in Fig. 9D. The staff assistant augments the capacity of the line manager. The arrangement has long been practised in industry, government, military and other organizations for providing managers with assistants and expert services.

In the case studies, there was an example in firm **Q**, where Associate Partner E assisted the Managing Partner A. The draughtsman in firm **M** assisting the chief engineer in analysing potential customers' requirements was working in a similar way.

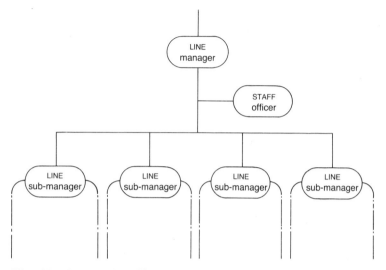

Fig. 9D. Line-and-staff principle

In Fig. 9D, the staff officer could assist the line manager in coordinating work for a project divided among the groups under two or more of the line sub-managers.

The important principle in this is that the staff role is not an additional level in the hierarchy. The principle applies whether there is one staff person or a department providing a service to a line manager. They are an extension of the line manager's capacity. Their formal authority to give instructions to the sub-managers can come only from the line manager. If they are respected and have knowledge of the facts of a project, their influence can be great.

Like anything logical, the line-and-staff arrangement works when all concerned understand it. In times of conflict or uncertainty it may be limited in its effectiveness because the line members of the hierarchy are in formal control of the resources.

Evolution of a system

Figure 9E illustrates stages in the development of a system of project coordination roles in a manufacturing firm. Production was in four over-lapping stages: conceptual design; development and detailed design; manufacturing and assembly; and performance

COORDINATION AND CONTROL

testing. Departments (in this firm called 'units') specialized in a stage of work for a project.

The firm was founded by the manager to use his special expertise to develop products for one customer. The customer was large. Because of their concern for the performance and delivery of the products, the customer's engineers planned and monitored the firm's work in detail. In effect, it employed the firm as if it owned it.

Decrease in that customer's orders led the manager to try to obtain work from others. Because design depended on development, potential customers demanded evidence that promised delivery times would be achieved. The manager therefore appointed a planner, in a Staff role to assist him as shown in Fig. 9E(b). This was not welcomed by some of the unit supervisors, because the new role was not defined.

Orders were obtained, but products were delivered late. To overcome this, an important customer insisted that the firm should add the project leader* shown in Fig. 9E(c). This was also in effect a Staff role of assisting the manager. It was also not understood by the supervisors, and the relationship with the planner was uncertain.

Discussions between these seven people eventually achieved agreement between them on an information system shown in Fig. 9E(d). The words used in this were chosen to distinguish between their Line authority as seen by the four supervisors and the Staff services to be provided by the planner and the project leader. Important in this was agreement that the project leader should monitor the progress of work at every stage of the projects, and with this information state the target dates best for meeting the firm's policy on priorities between customers. Agreement on this formal system for planning and monitoring led to the seven becoming a team which would then review and change relationships as the needs continued to change.

This case includes detail particular to the firm's history and type of work. It indicates a general lesson that new roles need to be agreed between all concerned before being added to a system.

---

*This neutral title, rather than the particular one given it in the firm, is used here for the role.

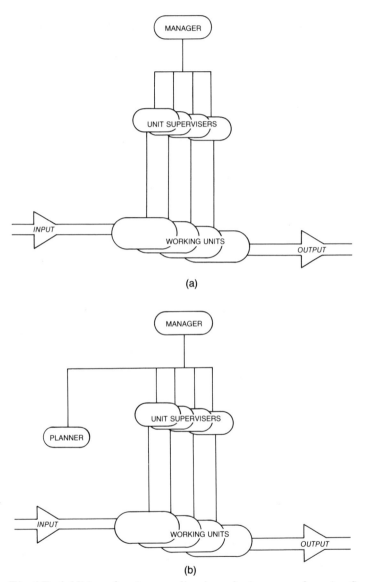

Fig. 9E. Addition of project coordination roles in a manufacturing firm: (a) original system of management; (b) addition of planner; (c) addition of Project Leader; (d) information flow needed

COORDINATION AND CONTROL

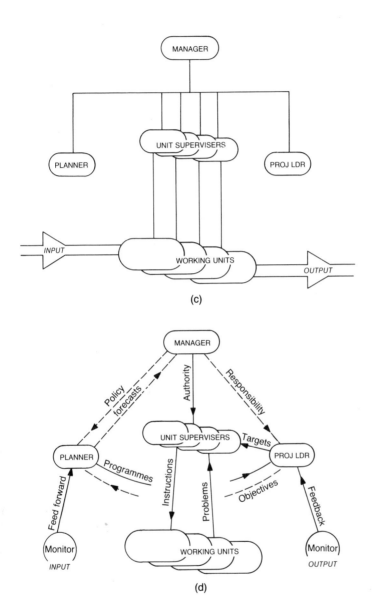

Fig. 9E. (continued)

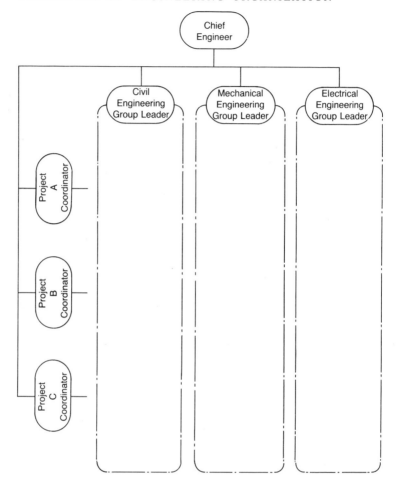

Fig. 9F. Matrix organization: specialism groups and project coordinators

Matrix systems

Cases in **Q** and **R** included instances of more complex systems based on what is called a *matrix* of authority.

Figure 9F illustrates the principle. In this simple example, each project coordinator* has the task of coordinating the members of

---

*The title 'Project Engineer' often means this role, but practice varies widely.

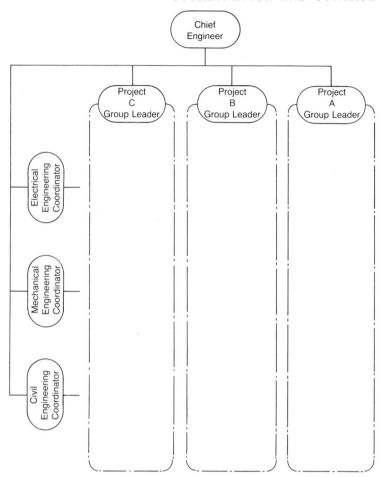

Fig. 9G. Matrix organization: project groups and specialism coordinators

the three specialist engineering groups who are working on their project. The engineering groups are the columns of resources. The project coordinators are the rows in the matrix.

For a larger number of projects, the set of coordinators could form a group under a chief projects manager.

An alternative for the same number of projects and size of department is that the resources are dedicated to each project so

PRINCIPLES OF ENGINEERING ORGANIZATION

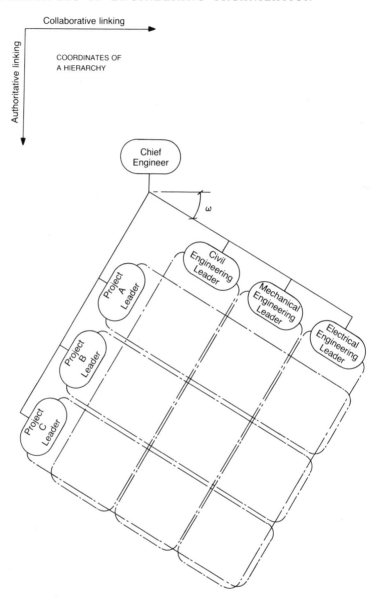

Fig. 9H. A matrix system of joint responsibility

long as needed, forming project task forces, with specialist co-ordinators as shown in Fig. 9F.

In examples such as indicated in Fig. 9G, the coordinators have the tasks of providing specialist advice to the project groups, coordinating related work, and planning and monitoring the allocation of resources to them.

In matrix systems, the project and specialist leaders should theoretically all influence decisions. To simplify this, some firms have stated that the project leaders in such systems are responsible for *quantitative* decisions affecting the cost and programme of the projects, whereas the specialist managers are responsible for *qualitative* standards in allocating people and other resources to each project. To show such a division of responsibilities the diagram can be drawn on the skew as in Fig. 9H.

In Fig. 9H the matrix is skewed because the convention in organizational diagrams is that the line of authority is vertical. Coordination is thus horizontal. If so, the angle w indicates the relative influence of project and specialist responsibilities for achieving results.

This system can be effective if the allocation of specialist people and other resources is planned by the specialist and project leaders jointly, and the group leaders become consultants to members of their groups once they are allocated to a project.

Changes in the value of w during the work for a project would be typical, depending on the relative importance of innovation and control stage by stage.

Relative responsibility

From these and other published examples, an ascending scale of at least nine degrees of relative project responsibility can be defined.

- Chaser: the minimum project role — a person sent to report on problems after they have become apparent
- Monitor: an observer of progress and problems, reporting these to specialist or project managers
- Coordinator: as above, but expected to anticipate problems without authority over any work for the project

- Planner: a provider of planning advice and services, but with no control over project decisions
- Administrator: a coordinator with responsibility for communications between departments or between firms
- Engineer: an initiator of the project, possibly the systems designer, but otherwise a coordinator as above
- Controller: an administrator with control over internal or contract payments
- Manager: combining all or most of the above, assisted by a team except on a smaller project
- Director: the maximum project role — a manager executively in control of every person employed on a project

The last of these categories consists of complete grouping into a project 'task force', the arrangement in a firm that is likely only for unique and urgent projects. Much more common, but unfortunately more complex, is the tendency to develop from employing the project chaser towards employing project managers sharing authority.

As yet, there is little evidence on how and why the coordination of roles can be successful. They can be most difficult to make effective when first added to a system. In studying such roles, C. Argyris observed that the position is one of power based on the best knowledge of the state and the needs of a project. If there is no policy for his project, the project coordinator must create one. If there is a lack of decisions, he must make them. He is the 'champion' for the project. One requirement in such a role is the wide knowledge that may lead to solutions to novel problems. Whatever the formal division of authority, the result should relieve the pressures on the specialist managers.

Matrix management or internal contracts?

Observations of matrix systems indicate that there can be problems in them between project managers and the heads of specialist groups about the allocation of resources to a project and the quality, cost and timing of the work to be done by them. The heads of the specialist groups should earlier have agreed on specifications, budgets and programmes for every project, but may

COORDINATION AND CONTROL

have done so some time before a project starts and then only in sufficient detail to get a budget or a contract to proceed. This may not ensure that adequate resources are available when a project calls for them.

One means of avoiding most or all such problems is to treat each specialist group's work for a project as contractual. If it were to be purchased by contract from another firm there would normally be a prior process of investigating the potential supplier's capacity and understanding of the work required, followed by an invitation to offer to do it for a price and specified quality and delivery. Procedures for progress reporting, inspection, changes and resolving problems would also be agreed before they might be needed. The same are in effect needed within organizations, not through legally enforceable documents but by written or oral definitions of what is expected of others, rather than the assumption that this is understood and agreed.

Three-dimensional matrix

Matrix systems for large projects can be three-dimensional, for instance with these three arms

- groups who design classes of sub-systems for projects: e.g. one designing transmission systems, and another designing engines for a series of vehicles
- groups specializing in expertise, providing a service to the above groups, for instance metallurgists and vibration experts
- project coordinators or teams.

Numbers

A survey of the numbers and work of people employed in some engineering departments indicated that the number employed on coordination is not related linearly to the number of people being coordinated. The form of the results obtained is indicated in Fig. 9I.

The independent variable D in Fig. 9I is the average number of designers per project in a department, the word designers being used to include the engineers, draughtsmen and technicians employed on all stages of design and the supporting development. The total of these in a firm is divided by the number of projects in hand. The dependent variable C depicted is the proportion of

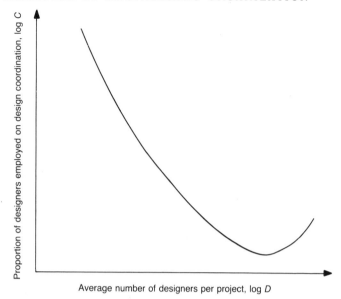

Fig. 9I. Proportion of designers engaged in coordination

designers in a department employed on the coordination of design (including development). The numbers for this were obtained from 22 firms, varying in size and in type of work.

The survey does not show what number of people *should* be employed per project or what proportion of them should be co-ordinators. The first should depend on the objectives and programme of a project, as the optimization of the time and cost of the whole project is likely to override optimization of any one stage alone. The consequential coordinating numbers per project could be analysed in a firm in order to make comparisons between projects, and with other firms.

The number of people employed on a project usually changes greatly from stage to stage in the work. These results indicate that the coordinating system should also change. It needs regular attention by managers, to anticipate possible future needs and new problems.

Further reading
Morris P. W. G., *Managing project interfaces*, Major Projects Association, Oxford, 1989, Technical Paper 7.

10 Final comments

Principles, practice and people
The chapters in this book proceed from the general to the particular, following the principle that the surrounding relationships of customers and suppliers should be considered before the choices within a firm. Designing a system of organization should correspond to the order of decisions in designing a project. Reviews of a system should proceed likewise. The external links needed with the demand for the firm's work and with others contributing to a project should be the starting point for deciding the internal system. Choices which are logical can then be made to meet objectives and the uncertainty of information.

Analysis and review of needs rather than tradition and habits are required. Firms should not discard their experience; they should analyse, modify and add to it. For instance, the division of design into stages established in some of the cases is inherently less suited to conditions of greater uncertainty, as decisions have to be recycled rather than handed on as final. On the other hand, combining it with geographical separation may make people plan better, discourage changes of mind, and, where needed, help achieve security of information.

Personal relationships at work are most likely to be satisfactory if based on interdependence which is accepted as logical for survival and success. The relationships between the people working on a project should be planned and agreed with them, and reviewed with them regularly. This is a recurrent lesson of studies of the causes of project successes and failures. Hence the earlier comment that managers should analyse *how* project problems and technical questions arise and *how* the decisions on them are made. Attempts to overcome personal conflicts are unlikely to be effective if systems are imposed which are thought to be wrong by the people in them. Some expedients are inevitable in practice, particularly when decisions appear to be urgent. These should not become habitual.

Projects and organizational change

In most if not all engineering firms, survival and success depend on a range of projects varying from the large and novel to minor alterations of existing investments. Every one of these disturbs the status quo, if only a little. This inherent characteristic of engineering does not fit with the bureaucratic principle that decisions are precedents which establish rules for solving a category of problem. Projects only establish the precedent that categories of problems will change.

An engineering organization therefore has to be open to the inherent need to change for the cycle of work for each project, requiring increasingly to achieve first a fluid stage of considering ideas, then a probabilistic stage of defining commitments, and finally the most detailed stage of integrating physical actions.

Innovations in projects and in the system for achieving these innovations are interdependent. A system of roles must be designed first, in order to offer these roles to people, but once in them they will have to evolve changes in their roles and relationships in order to adapt to unpredictable problems. If there is sufficient slack, the system can thus 'co-opt the informal organization' (D. W. Conrath) as it evolves.

Larger and more organic systems

Economic pressures and technological advances have led to larger and more complex organizations. There is no obvious reason why this should reverse and so make working relationships simpler.

The more complex links needed are not necessarily achieved by more formal divisions of work. The trends vary from firm to firm, but organizational habits derived from factory ways of dividing predictable and repetitive manufacturing operations cannot sensibly be assumed to be appropriate to the abstract and probabilistic process of engineering decisions. The evidence indicates that the concepts of hierarchies of authority are limited in value, and that in reality more complex systems evolve in order to be open to changes in demand and to meet unpredictable problems.

Data systems

The increasingly greater use of computer-based data processing and simulation systems seems unlikely to end the problems of uncertainty in design decisions and complexity in engineering organizations. Perhaps the opposite is likely, as these tools are being used to increase the complexity of projects.

They are being used to store, adapt and transmit design details. This is likely to have a standardizing effect. Feedback during design and feedforward of experience to new projects should improve. Computer systems can be programmed to process the predictable decisions and be linked directly with the machines used in manufacturing, plant installation, construction, testing, commissioning, operation and dismantling. This automation is consistent with the pattern of all industrial evolution. The decisions on the unforeseen problems so far remain dependent on people.

Uncertainty and safety

Greater uncertainty in predicting demands requires more investment to design alternative proposals and analyse their risks ahead of commitment. Otherwise time is lost in redesigning later.

Failures of engineering products are relatively few. Some that do occur are dramatic and tragic, as technological advances have made possible larger and larger concentrations of energy in systems and materials. Specialization disperses the knowledge which might anticipate these failures. Decisions in design are increasingly dependent on compounding the comprehension of people whose cognitive limits cannot stretch across all the complexities of a project. Systems of organization to relate their decisions are therefore increasingly important, but in them there must be the slack and the will to use it to perceive the potential problems of safety and economy.

An increasing proportion of the eventual costs of projects should therefore be spent in exploring potential problems and risks before commitment to a cost, specification and programme.

Coordination

The cases demonstrate that the principle of a hierarchy of authority is commonly used to make the critical decisions to commit resources to a project, but that complicating additions are made to link specialists into projects.

The employment of people in specialist groups has been the dominant trend in all industries. It tends to limit their outlook and their abilities. It is the organizational habit that is likely to breed closed minds, certain of their solutions. An alternative is to employ some specialists in 'Staff' roles, to advise on problems and to review design decisions. They may best operate informally and cooperatively, in order to avoid being the conservative influence that is likely if they are to share the risks of decisions but not the credit for successes.

The addition of project roles to a system of specializations by subject or stage of work can establish responsibilities to define problems in terms of the firm's objectives. This should make explicit any conflicts between the development of specialisms and the usually more immediate but transient demand to employ these specialisms on each project when required.

The cases show how the authority for decisions in an engineering organization can be defined according to the relative importance and complexity of its projects. There can be a matrix of roles in which decisions are made jointly between the temporary project coordinators and the continuing specialists, or between the project teams and specialism coordinators. The greater the uncertainty in decisions, the more such systems may be needed. The coordinators need not be permanent. Specialists can take turns at this role. More important, any of these complications to systems require a sophisticated understanding of the change from the simple principles of organization trees.

Similar linking is also needed between sets of firms contributing their specialisms to a project, especially in a joint venture or consortium for temporary collaboration between firms with disparate interests. A continuing trend in sectors of industries is for companies to amalgamate. These temporary and permanent moves are expected to achieve economy of scale in the production of goods and services. They do not usually link firms in the ways required for projects. This has to be arranged for each project and increasingly requires systems to link their related decisions from the start of entering into a new commitment.

Project teams

For a special project the formation of a dedicated team or 'task force' can achieve links between organizations or groups in an organization. The demand for this can come from a customer, because of his concern about a supplier's arrangements to complete a commitment. The cost to the supplier is obvious; the advantages to the supplier may not be obvious. It may therefore have to be imposed by the customer.

It is logical for every representative in a project team to have the authority to commit his parent organization, whether based on formal position or on expertise (C. Argyris). Using such a system, the decisions can be made collaboratively at the most knowledgeable level in organizations, which again is a change from some traditions and concepts.

The designation of a separate project team involving all those affected can also be a means of achieving an organizational transition, utilizing people's interest in the technical novelties or other features of a project to draw them into a changed system.

Case studies

Many of the preceding statements have been in relative terms, commenting on the pressures on firms to change because of the 'increasing complexity' of projects.

Relative statements are unavoidable so long as the variables affecting systems of organization cannot be measured. Perhaps some of the variables are not measurable, particularly those dependent on human behaviour. Others, as yet, may be obscure because of our lack of understanding. Comparisons of systems help, by contrasting several firms to try to see how differences in demands appear to correlate with differences in systems, and then by studying a firm more at length to try to learn from its responses to changes. The cases described are not complete examples, and are not necessarily exemplary. They illustrate applications of the principles reviewed earlier, but are not necessarily models for copying. They provide a basis for comparisons when studying the choices possible in meeting the demands in a firm.

Flow diagrams

Organigrams and similar ways of defining authority and responsibility are used quite commonly in firms. The case studies show the value of using flow diagrams with them to analyse and to plan relationships. All the detail of communications between people in an organization cannot be readily shown, but using them to define the sequence and nature of the critical communications provides a basis for discerning the causes of problems and for starting to remedy them.

More general diagrams were used in chapters 4 and 5 to illustrate contractual relationships between firms. Understanding of these relationships can also be improved by using a diagram showing the flow of decisions and other information. Whether the people on a project are employed in firms linked contractually or all in one firm and linked managerially, a diagram such as the example in Fig. 10A could help all to understand and to improve their relationships.

Figure 10A illustrates critical links between the parties to a large construction project. It was a result of case studies which demonstrated the value to project promoters of appointing an authoritative project director. This type of diagram could be used to define or review the relationships between all parties to projects of any type or size.

Variety and further studies

It has been argued that engineering organizations have to be flexible to meet changes. The demands on them vary. The system suitable for a firm is not likely to be suitable for its customers or for its suppliers. The specification of a standard system has not been our objective. Nor have we discovered one. A great variety of systems can be found in use. Most or all may be logical, but those which have been studied show that the logic of what they need or what they have may not be apparent to the members of an organization. More investigations to compare theory with practice could therefore aid the managers and all the members of organizations, as well as add to knowledge.

Relatively few real systems have yet been investigated. The ones that have been show that there are logical choices in designing systems and they illustrate the application of these to meet contrasting demands. Many doubts and problems remain for further investigation. Engineers and managers no doubt have various preferences, experience and opinions on these and other questions. More evidence on what is successful and why could be helpful. As complexity in engineering is unlikely to decrease, at least in the immediate future, we should catch up in learning from how the problems are being solved and apply the theories of how they may be better understood.

FINAL COMMENTS

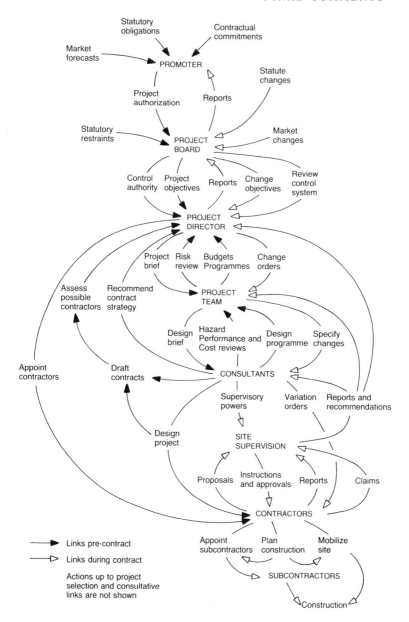

Fig. 10A. Responsibilities for a large construction project (Ninos and Wearne)

Investigations might show that the larger the organization, the less attention is given in it to detail. The pressures and problems of scale and complexity vary greatly from firm to firm, but there are these and other questions which can affect any choice of system. Firms make their decisions largely in isolation from each other, even in the public services. Much therefore might be learnt from more investigations to analyse practice.

The 'deep-rooted beliefs' of traditions in firms may also be imperfect and some may be wrong. They do lack validation. At present they seem to be fertilized by familiarity rather than turned over selectively. Theory provides the means of analysing practice and clarifying possible choices in systems. Practice provides the means of testing and developing the theory. The two need to advance together.

Engineering services are increasingly vital to the survival of communities, in cities and in agriculture. Food and pharmaceuticals manufacturing, power, public health, water supply and transport are major examples. Engineering is also an important sector of employment. Its systems must be adaptable to changing requirements. They must try to make the best use of the ideas and efforts of the people they employ.

Appendix A. Charts and diagrammatic conventions

This appendix describes the type of charts and conventions used in the book to represent and analyse systems of organization.

Relationships between activities

The relationships between activities in industry can be complex but any two activities can be related in only two ways: either the two run in *parallel* or they run in *series*. These alternatives are illustrated in Fig. A1.

These elementary relationships can occur many times in a more complicated system. Activities can overlap, and work may be recycled between two or more activities which are then *reciprocally* dependent.

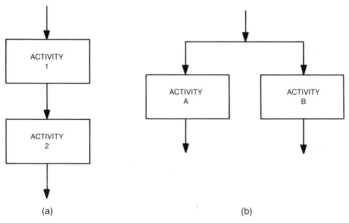

Fig. A1. Elementary relationships between activities: (a) vertical relationship of activities in series; (b) horizontal independence of activities in parallel

135

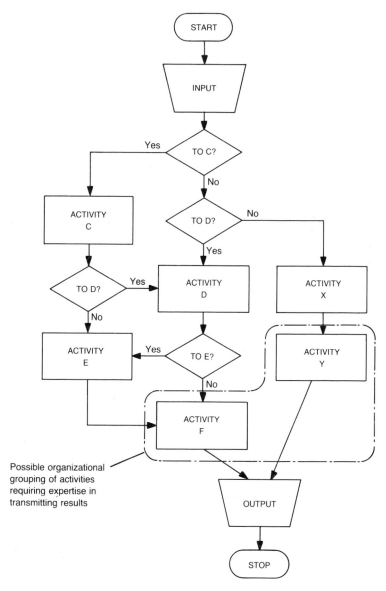

Fig. A2. Flow chart

APPENDIX A

Flow charts

Flow charts such as the simple example shown in Fig. A2 can be used to look at and design flows of information or of materials. They are useful when studying flow in manufacturing, construction or information systems.

The conventional signs of box shapes used in these flow charts follow practice in systems analysis and computer programming. These conventions are particularly suitable as information systems increasingly use computers.

In the conventional signs, the rectangular boxes denote the activities which take time. The diamond-shaped boxes at junctions denote decision points. Flow along the arrows and through these junctions is taken as automatic and instantaneous compared with the time taken in any activity. The length of an arrow does not imply time taken, and is a result only of the layout of boxes. The chart is thus similar to an 'activity-on-node' type of network used in project planning.

The relationships shown on a flow chart may be studied in several ways when considering a system of organization. The example in Fig. A2 shows that some sequences of work consist of more activities than others, the sequence C–D–E–F including the largest number of activities and D–F and X–Y being two which are more direct.

The time taken through one sequence relative to another will depend on time taken on the activities in it, but the chart demonstrates the need to consider the capacity for each activity to balance flows through the system. The chart makes this and other details obvious. In Fig. A2 the activities X and Y are dependent on each other but not on others in the system. Logically, the people employed on these two activities might therefore be grouped together.

Those employed on C, D and X might also be grouped together, as they could share expertise in interpreting incoming flow. E is related to three others but is unique in having connections to activities only within the system shown in the chart.

To make such charts complete, all connections must show the direction of flow. The set of conventional signs used here are shown in Fig. A3. These can be used to show more features of relationships.

The first example shown in Fig. A1 is redrawn in Fig. A4 with additional detail. Two of the additions shown are feedback loops

PRINCIPLES OF ENGINEERING ORGANIZATION

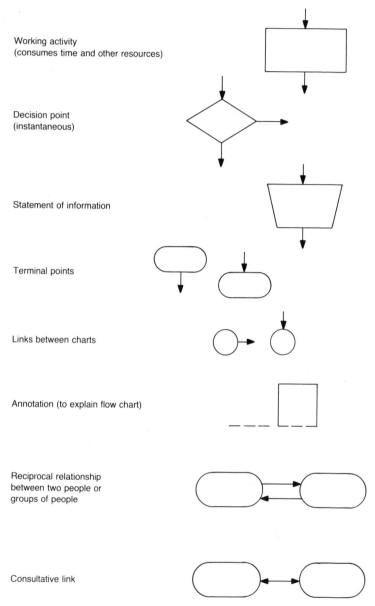

Fig. A3. Conventional signs for flow charts

APPENDIX A

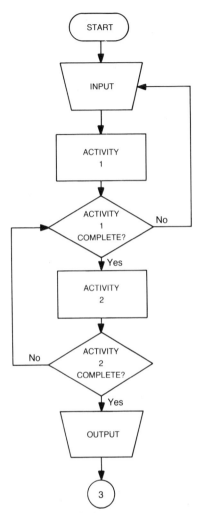

Fig. A4. Serial activities with feedback

of information. These feedbacks could be either information from monitoring an activity to decide whether to change it until the result was acceptable for passing on, or for reporting that a stage of work was clear for further input. In this example, a display of the result after the second activity has also been added.

Flow charts can be used to show the relationships of activities shared between firms or between groups of people in a firm, when analysing what appears to be occurring in the system being studied or when trying to predict the effects of changes or a new system.

The complications in depicting systems of organization come where the logical relationships of the activities essential for a project cross the boundaries of organizations sharing the work. These boundaries can be shown on flow charts. In Fig. A2, a broken line is used to show the possible boundary of grouping together the people in a firm who are working on activities F and Y.

Vertical, horizontal and lateral relationships

Vertical is the word used conventionally to mean a system in which a sequence of activities within a firm is divided amongst specialist groups working in series, as shown in Fig. A1(a). Vertical integration is therefore the term for the process of combining some firms which had been separate specialists working in series.

Horizontal or *lateral* are used to mean a firm that consists of several subsidiary businesses which operate independently in their own specialism, as in Fig. A1(b). Horizontal integration is therefore the process of combining them, for instance to share information, processes or other resources.

The origin of these conventional terms was in describing divisions of work in manufacturing.

Management levels: middle, top, bottom

Also widely understood is the convention that the greatest authority is at the *top* of an organization, and is divided and subdivided through middle managers down to the supervisors of members at the bottom. This is commonly used in describing systems of management in firms. The head of the organization is naturally visualized as at the top of the body.

This hierarchical language is used for steps of divisions of authority and responsibility. What are described as the top and the bottom in a system of organization are not necessarily related to a vertical or a horizontal integration of activities, as discussed in analysing the case studies. The organigrams in chapters 7 and 9 follow this convention.

Time dimension

No one direction in diagrams can be used consistently to imply

APPENDIX A

the passing of time, as the sequences of activities and flows of information for a project may pass in any direction relative to a hierarchy of authority and the relationships between firms.

A time-scale starting at the left of each diagram is implied in Figs 2A–2D along the horizontal axis. This is the convention in bar charts used in planning where time is the independent variable.

The flow charts in chapter 7 use the convention of starting at the top of the page, centrally or towards the left. The sequence therefore proceeds generally downwards, but there is no rule in this, as the important action is to arrange the diagram so as to show all relationships. Flows also usually include some recycling, as indicated in Fig. 2F, and this cannot be shown by lines in only one direction in a diagram. Flow charts are used to show logical relationships, and no dimension of time is possible or directly relevant until they are drawn.

Appendix B. Number and scope of contracts

This appendix draws on experience and publications on the choices in contract responsibilities for the design and supply of machines, structures and engineering services.

In engineering and construction contracts the supplier of goods or services is usually called 'the contractor'. Various words are used to mean the purchasing firm. To be consistent the word 'promoter' is used here.

One comprehensive contract
A comprehensive contract for the whole project has the following potential organizational advantages to the promoter:

- The contractor has to manage all the relationships between design, procurement, subcontractors, etc.
- The promoter has only the one contractor to deal with, so his project team should be relatively simple in organization and smaller than required for other types of contract.
- The promoter's staff can concentrate on project objectives. Their minimum involvement in design may result in fewer changes.
- The contractor has the greatest scope for being efficient, and therefore economical and profitable.
- The contractor should be able to plan design and procurement to suit construction and installation, and so offer shorter programmes.

A comprehensive contract has the following potential disadvantages compared with other types of contract:

- The relationships between design, manufacturing,

APPENDIX B

subcontractors, installation and commissioning remain the same. Management of them is shifted to the contractor and away from the promoter's control.
- Much of the work is usually subcontracted, so the promoter has limited or only indirect ability to assess and influence whether or not it will be completed correctly and on time.
- Few contractors may have the financial strength or be able and willing to manage all of a large project, so that the choice of contractor may be limited.
- Many specialist engineering companies prefer to work directly for the ultimate user of their products and services. If employed as subcontractors, they may not be so well motivated to perform.

A comprehensive contract is common for new process plant projects, industrial power stations, office blocks, warehouses and other non-novel buildings for commercial use, and for many projects in developing countries.

A series of contracts

The following are the potential advantages to a promoter of proceeding by stages in a vertically related sequence of two or more contracts:

- Starting with a contract for design only should make it possible for most design uncertainties and changes to be settled before they affect the contracts for the more expensive physical activities of manufacturing machines, installation, etc.
- The design decisions which most affect project cost are separated from the influence of the contractors.
- A newly formed promoter's team has time to prepare to enter into the larger value contracts.
- A contractor can be chosen as best for each stage of work.
- If a design contractor is not the contractor for the consequent manufacturing or construction work his staff can be employed as its supervisors or advisers.
- The end of each stage is an opportunity to review new information and to change, confirm or stop the project.
- The promoter can vary the speed of continuing with the project.

A sequence of contracts has the following potential disadvantages:

- The promoter has to plan and manage the interactions between the sequence of contracts and has to take the risk that errors by one contractor will cause claims on the promoter from one or more other contractors.
- The promoter has no firm indication of the final cost of the project until the final contracts have been made.
- No party can plan far ahead.
- The time needed at each stage to select a contractor, start a new contract and establish procedures between all parties could delay completion of the project.
- The promoter's and contractors' staff tend to concentrate on their stage of work and may not be directly responsible for meeting the final objectives of the project. They may also get limited or no feedback of the ultimate results of their contribution.

The division of design and construction into separate contracts is traditional for civil engineering and building projects in Britain and other countries. The design and construction of some large risky and novel projects are divided into more stages, for instance large offshore platforms.

Parallel contracts

The following are the potential advantages of dividing work among horizontally related contractors in parallel:

- A contractor can be chosen as best in expertise or resources for the one particular type of work.
- Confidential information is limited to the promoter and the contractors involved.

Such a division of work has the following potential disadvantage:

- The promoter has to plan and manage the interactions between the parallel contracts and to bear the risks that actions by one contractor will cause claims from the other contractors.

Parallel contracts are very common, for instance the maintenance and supply contracts placed by manufacturers and the sub-contracts placed by main contractors. Parallel contracts are usual in some

APPENDIX B

countries for each type of construction work for civil engineering and building projects.

Further reading

Horgan M. O'C. and Roulston F. R., *Project control of engineering contracts*, Spon, 1988.

Loraine R. K., *Construction management in developing countries*, Thomas Telford, London, 1991.

Marsh P. D. V., *Contracting for engineering and construction projects*, Gower Press, 1989, 3rd edn.

Smith N. J. and Wearne S. H., *Construction contract arrangements in EC countries*, European Construction Institute, Loughborough, 1993.

Appendix C. Conduct of project completion reviews

This appendix draws on experience in various industries of reviewing completed projects for the purpose of enabling firms and individuals to define the lessons they should apply to further ones.

Principles
A review of a completed project should be planned and conducted so as to assess what went well, not just concentrated on problems, so that the conclusions can list what methods and processes were satisfactory as well as state which need changing.

Recommendations based only on problems encountered can be unbalanced and lead to changes which affect what were successful methods, etc. A review should therefore look at all the major decisions stage by stage in the project, and explain the expected and unexpected results.

Rules
To achieve its objectives a review should:

- involve all who were involved in decisions, preferably by their meeting together so as to agree facts, to balance conclusions and to enable all to learn
- pursue facts, not blame, so as to define what could have been known at the time of a decision
- not rely on recollections: the project records should be summarized in preparation for the review so as to avoid all possible doubts
- not expect perfection: past decisions should be assessed as to whether or not they were reasonable at the time, remembering

APPENDIX C

that perfection would bankrupt a project and that taking risks is good management
- not pursue universal answers: rarely is there a unique best answer to any decision, so a review should concentrate on why choices were successful or unsuccessful in the circumstances at the time.

Procedure

1. Announce that a review is proposed to define collective and individual lessons to apply to future projects, not to discuss blame, so as to dispel anxiety. Support for this purpose should be demonstrated by all senior managers.
2. Ask every group (larger projects) or every individual (smaller projects) who worked on any stage of the project to prepare collective summaries of decisions and other facts, as preparation for a review meeting.
3. Get all the above or representatives of them to the meeting, if possible, and begin by restating the purpose and then asking them in turn to state but *not* then to discuss:

   - what each, in his opinion, considers to have been the one most successful feature of the project
   - what each considers to have been the one greatest problem.

   List both the above as reminders to guide discussion and record them at the start of a report on the conclusions.
4. Then review the project decisions stage by stage through the project's definition, engineering, completion and handover, using the following check list for each stage:

   - What was known at the start?
   - How was this information used?
   - What was the plan?
   - How was it agreed?
   - What problems were expected?
   - What actions were taken?
   - What were the successful results?
   - What were the problems?
   - What could have been known at the time to avoid or reduce these problems?
   - What are the lessons?

The people involved at each stage should be invited to be the first to answer the above.

Output

The results of a review can be conveyed in all the following ways:
- A project review report, summarizing facts and lessons, providing the long-term record. It should also record any dissent from the lessons, with the reasons.
- **GOOD IDEAS** sheets, one per idea, for short-term or cycled display in appropriate offices
- **DON'T DO IT AGAIN** posters, also for short-term or cycled display
- check lists for:

  corporate planners
  chief engineer
  design leaders
  project managers
  project engineers.

All these and other means should be considered for conveying the results of a review.

General comments

People tend to vary in their recollections and perception of events, and so differ about what could have been done about problems and about achieving happier results. To be of benefit, a project review therefore needs to be a team task. People also tend more readily to recall problems and the difficulties of working with others to control and solve them. We therefore emphasize the potential value of a review as an opportunity to agree conclusions on all the lessons, not only on what needs changing, but also on what went well and why.

A review costs money, mainly in requiring time from the people responsible to analyse their part in a project and to discuss the results. The gain should be that individual views are more accurately based, broadened by attention to the successes as well as the problems, and the lessons defined for others to be able to use. Recollections alone, informal discussions or unprepared reviews are unlikely to do this. A review should therefore be

APPENDIX C

planned, budgeted and controlled to achieve its objectives — as in effect a mini-project.

Attention in a review should be given not only to *what* was decided and the results, but also to *how* information was obtained, used and communicated, i.e. to the processes of project management. To many involved a project tends to be thought of as unique, and the events particular to the product of hardware and software, the people involved, the problems and decisions. In their combination and content these can be unique, and the lessons for other projects not therefore obvious. Common to all projects is the process of moving from general objectives to technical detail, the increasing use of resources, risks in achieving plans, and a transition from project execution to use or operation of the resulting equipment or service. Future projects can therefore benefit from knowing how this process was planned, budgeted, organized, led and controlled to suit the size, complexity, novelty, risks, urgency and other priorities of each project.

Development of skills and methods for project management depends on learning from the application of ideas and techniques to real projects. Industry is therefore its own project management laboratory, and learning achieved by time and discipline in recording and analysing these decisions and their results.

Ideally every project should provide a contribution to this by being the subject of a post-completion review.

Reference

Corrie R. K. (ed.), *Project evaluation*, Thomas Telford, London, 1991.

Appendix D. Bibliography — Sources

The following is a list of publications drawn on in the preparation of the first and second editions of this book.

Allen G. C., *The structure of industry in Britain*, Longmans, 1970.
Argyris C., Today's problems with tomorrow's organizations, *J. Mgmt Studies*, 1967, **4**, no. 1, 31–35.
Barnes L. B., *Organization systems and engineering groups*, Harvard University Press, 1960.
Beer S., *The brain of the firm*, Allen Lane and Penguin, London, 1972.
Brech E. F. L., *Organization, the framework of management*, Longmans, 1965, 2nd edn.
Burns T. and Stalker G. M., *The management of innovation*, Tavistock Publications, 1966, 2nd edn.
Checkland S. G., *The rise of industrial society in England*, Longmans, 1964.
Cohen J. and Christensen I., *Information and choice*, Oliver & Boyd, 1970.
Conrath D. W., The role of the informal organization in decision making in research and development, *IEEE Trans. Engng Mgmt*, 1968, **EM-15**, no. 3, 109–119.
Dill D. D. and Pearson A. W., The effectiveness of project managers: Implications of a political model of influence, *IEEE Trans. Engng Mgmt*, 1984, **EM-31**, no. 3, Aug., 138–146.
Dill D. D. and Pearson A. W., The self-designing organization: Structure, learning and the management of technical professionals, *Int. Conf. on the Management of Engineering and Technology*, Portland, Oregon, 1991, 33–36.
Papers on *Flat, Flexible Organizational Structures and Projects*, 10th Internet Congress, Vienna, 1990.
Johnson J. K., Organization structures and the development project planning sequence, *Public Administration & Development*, 1984, **4**, 111–131.

Kingdon D. R., *Matrix organization*, Tavistock Publications, 1973.
Langrish J. et al., *Wealth from knowledge*, Macmillan, 1972.
Lawrence P. R. and Lorsch J. W., *Organization and environment*, Harvard, 1967.
Lock D. (ed.), *Project management handbook*, Gower Press, 1987.
Marsh P. D. V., *Contracting for engineering and construction projects*, Gower Press, 1989, 3rd edn.
Martin A. S., *Can you manage?*, Municipal Publications, 1981.
Martin A. S., Management organisation in a district council's technical services department, *Planning, Transportation, Research and Computation* Ann. Meeting, Warwick, 1974.
Middleton C. J., How to set up a project organization, *Harvard Business Review*, 1967, Mar.–Apr., 73–82.
Miller E. J. and Rice A. K., *Systems of organization*, Tavistock Press, 1967.
Miller R. J., *Consulting Engr*, 1970, Oct., 43–47, and 1970, Nov., 61–63.
Morris P. W. G. and Hough G. H., *The anatomy of major projects*, Wiley, 1987.
Ninos G. E. and Wearne S. H., Control of projects during construction, *Proc. Instn Civ. Engrs*, Part 1, 1986, **80**, 931–943, and 1987, **82**, 859–869.
Nochur K. S. and Allen T. J., Do nominated boundary spanners become effective technological gatekeepers?, *IEEE Trans. Engng Mgmt*, 1992, **EM-39**, no. 3, Aug., 265–269.
Ramström D. and Rhenman, *A method of describing the development of an engineering project*, Swedish Institute of Administrative Research, Stockholm, 1965.
Slater Lewis J., *The commercial organization of factories*, Spon, 1896.
Stinchcombe A. L. and Heimer C. A., *Organization theory and project management — administering uncertainty in Norwegian offshore oil*, Norwegian University Press/Oxford University Press, 1985.
Taylor F. W., *Shop management*, Harper, 1911.
Tebay R., Atherton J. and Wearne S. H., Mechanical engineering design decisions, *Proc. Instn Mech. Engrs*, 1984, **198B**, 87–96.
Twiss B. C., *Managing technological innovation*, Longmans, 1980, 2nd edn.
Wearne S. H., Organisation of engineering design departments — pilot numerical comparisons. *Int J. Production Research*, 1970, **8**, no. 2, 149–168.
Woodward J., *Industrial organization — behaviour and control*, Oxford University Press, 1970.

About the Author

Stephen Wearne is a consultant and a senior research fellow with the Project Management Group at UMIST. He is the director of short courses and in-company training in project and contract management run by engineering institutions in the UK and in other countries. He was originally a factory apprentice and 'sandwich' student. After training in water power engineering he worked on the design, economic planning and coordination of projects in Spain, Scotland and South America. In 1957 he joined turnkey contractors, on construction and design and then contract and project management of projects in this country and Japan. In 1964 he moved into research and teaching on engineering management, first at the University of Manchester Institute of Science and Technology, and from 1973 to 1984 as Professor of Technological Management at the University of Bradford. His research has included studies of design and project organizations, joint venture and consortia organizations and risks, engineering and construction contracts, project control, plant commissioning teams, the lessons of major failures and their causes, and the managerial tasks of engineers in their careers. He was first Chairman of the UK Engineering Project Management Forum initiated in 1985 by the National Economic Development Office and the Institutions of Civil, Mechanical, Electrical and Chemical Engineers.